Emotional Blackmail

Also by Susan Forward

Betrayal of Innocence
Men Who Hate Women and the Women Who Love Them
Obsessive Love
Toxic Parents
Money Demons

Emotional Blackmail

WHEN THE PEOPLE IN YOUR LIFE USE FEAR, OBLIGATION AND GUILT TO MANIPULATE YOU

Susan Forward, Ph.D.
with Donna Frazier

A hardcover edition of this book was published in 1997 by HarperCollins Publishers.

HarperCollins books may be purchased for educational, business, or sales promotional use. For information please write: Special Markets Department, HarperCollins Publishers Inc., 10 East 53rd Street, New York, NY 10022.

First HarperPerennial edition published 1998..

Reprinted in Quill 2001.

Designed by Alma Hochhauser Orenstein

The Library of Congress has catalogued the hardcover edition as follows:

Forward, Susan..
Emotional blackmail : when the people in your life use fear, obligation, and quilt to manipulate you / by Susan Forward with Donna Frazier. — 1st ed.
p. cm.
ISBN 0-06-018757-3
1. Manipulative behavior. 2. Control (Psychology). I. Frazier, Donna. II. Title.
BF632.5.F67 1997
158.2—dc21 97-3399

ISBN 0-06-092897-2 (pbk.)

08 RRD 40 39 38 37

Contents

Introduction

I told my husband I was going to take a class one night a week and he went ballistic in that quiet way of his. "Do whatever you want, you always do anyway," he told me, "but don't expect me to be waiting for you when you get home. I'm always there for you; why can't you be there for me?" I knew his argument didn't make sense, but it made me feel so selfish. I asked for my registration fee back. —LIZ

I was planning to spend Christmas traveling with my wife, a vacation we'd been looking forward to for months. I called my mom to tell her the news that we'd finally gotten the tickets, and she almost started to cry. "But what about Christmas dinner?" she said. "You know everyone always gets together for the holiday. If you go on that trip instead of coming, you'll ruin the holiday for everyone. How can you do this to me? How many Christmases do you think I have left?" So of course I gave in. My wife's going to kill me, but I don't see how I could enjoy a vacation while I'm buried under all that guilt. —TOM

I went in to tell my boss that I had to have help or a more realistic deadline on a big project I'm doing. As soon as I mentioned that I really needed some relief, he started on me. "I know how much you want to get home to your family," he said, "but even though they miss you now, you know they'll appreciate that promotion we've been considering you for. We need a team player with real dedication for that job—that's what I thought you were. But go ahead. Spend more time with the kids. Just remember that if those are your priorities, we might have to reconsider our plans for you." I felt totally blindsided. Now I don't know what to do. —KIM

What's going on here? Why do certain people leave us thinking "I've lost again. I always give in. I didn't say what I was really feeling. Why can't I ever get my point across? How come I can never stand up for myself?" We know we've been had. We know we feel frustrated and resentful, and we know we've given up what we want to please someone else, but we just don't know what to do about it. Why is it that some people are able to emotionally overpower us, leaving us feeling defeated?

The people we're coming up against in these can't-win situations are skillful manipulators. They swathe us in a comforting intimacy when they get what they want, but they frequently wind up threatening us in order to get their way, or burying us under a load of guilt and self-reproach when they don't. It may seem as though they map out ways to get what they want from us, but often they're not even aware of what they're doing. In fact, many can appear sweet or long-suffering and not threatening at all.

Generally, it's one particular person—a partner, a parent, a sibling, a friend—who manipulates us so consistently that we seem to forget everything we know about being effective adults. Though we may be skilled and successful in other parts of our lives, with these people we feel bewildered, powerless. They've got us wrapped around their little fingers.

Take my client Sarah, a court reporter. Sarah, a vivacious brunette, has been seeing a builder named Frank for almost a year. A close couple in their 30s, they got along well—until the subject of marriage came up. Then, said Sarah, "his whole demeanor toward me changed. He seemed to want me to prove myself." It all became clear one weekend when Frank invited her up for a romantic weekend at his cabin in the mountains. "When we arrived, the place was full of tarps and paint cans, and he handed me a brush. I didn't know what else to do, so I painted." They worked, mostly in silence, all day, and when they finally sat down to rest, Frank pulled out a huge diamond engagement ring.

"I asked him, 'What's going on?'" Sarah said, "and he said he needed to know that I was a good sport, that I would pitch in and not expect him to do everything in the marriage." Of course, that wasn't the end of the story.

We set a date and everything, but we went up and down like a yo-yo. He kept giving me gifts, but he also kept testing me. If I didn't want to go take care of his sister's kids one weekend, he said I didn't have a strong sense of family and maybe we should think about calling off the wedding. Or if I talked about expanding my business, it meant I wasn't really committed to him. So of course I put that on hold. It went on and on, with me always giving in. But I kept telling myself what a great guy he was and that maybe he was scared of getting married and just needed to feel more secure with me.

Frank's threats were quiet, yet they were powerfully effective because they alternated with a closeness enticing enough to obscure what was really going on. And like most of us, Sarah kept coming back for more.

She gave in to Frank's manipulations because, in the moment, making him happy seemed to make sense—there was so much at stake. Like so many of us, Sarah felt resentful and frustrated at Frank's threats, but she justified her capitulation to them in the name of peace.

In such relationships, we keep our focus on the other person's needs at the expense of our own, and we relax into the temporary illusion of safety we've created for ourselves by giving in. We've avoided conflict, confrontation—and the chance of a healthy relationship.

Maddening interactions like these are among the most common causes of friction in almost every relationship, yet they're rarely identified and understood. Often these instances of manipulation get labeled *miscommunication.* We tell ourselves, "I'm operating from feelings and he's operating from intellect" or "She's just coming from a different mind-set." But in reality, the source of friction isn't in communication styles. It's more in one person getting his or her way at the expense of another. These are more than simple misunderstandings—they're power struggles.

Over the years I've searched for a way to describe these struggles and the troubling cycle of behavior they lead to, and I've found that people almost universally respond with a charge of recognition when I tell them that what we're talking about here is pure and simple blackmail—*emotional* blackmail.

I realize that the term *blackmail* is one that conjures up sinister images of criminals, fear and extortion. Certainly it's difficult to think of your husband, your parents, your boss, your siblings or your children in that context. Yet I've found that *blackmail* is the only term that accurately describes what's going on. The very sharpness of the word helps us pierce the denial and confusion that cloud so many relationships, and doing that brings us to clarity.

Let me reassure you: Just because there's emotional blackmail in a close relationship doesn't mean it's doomed. It simply means that we need to honestly acknowledge and correct the behavior that's causing us pain, putting these relationships back on a more solid foundation.

WHAT IS EMOTIONAL BLACKMAIL?

Emotional blackmail is a powerful form of manipulation in which people close to us threaten, either directly or indirectly, to punish us if we don't do what they want. At the heart of any kind of blackmail is one basic threat, which can be expressed in many different ways: *If you don't behave the way I want you to, you will suffer*. A criminal blackmailer might threaten to use knowledge about a person's past to ruin her reputation, or ask to be paid off in cash to hide a secret. Emotional blackmail hits closer to home. Emotional blackmailers know how much we value our relationship with them. They know our vulnerabilities. Often they know our deepest secrets. And no matter how much they care about us, when they fear they won't get their way, they use this intimate knowledge to shape the threats that give them the payoff they want: our compliance.

Knowing that we want love or approval, our blackmailers threaten to withhold it or take it away altogether, or make us feel we must earn it. For example, if you pride yourself on being generous and caring, the blackmailer might label you selfish or inconsiderate if you don't accede to his wishes. If you value money and security, the blackmailer might attach conditions to providing them or threaten to take them away. And if you believe the blackmailer, you could fall into a pattern of letting him control your decisions and behavior.

We get locked into a dance with blackmail, a dance with myriad steps, shapes and partners.

LOST IN THE FOG

How do so many smart, capable people find themselves groping to understand behavior that seems so obvious? One key reason is that our blackmailers make it nearly impossible to see how they're manipulating us, because they lay down a thick fog that obscures their actions. We'd fight back if we could, but they ensure that we literally can't see what is happening to us. I use fog as both a metaphor for the confusion blackmailers create in us and as a lens for burning it off. FOG is a shorthand way of referring to Fear, Obligation and Guilt, the tools of the blackmailer's trade. Blackmailers pump an engulfing FOG into their relationships, ensuring that we will feel afraid to cross them, obligated to give them their way and terribly guilty if we don't.

Because it's so tough to cut through this FOG to recognize emotional blackmail when it's happening to you—or even in retrospect—I've devised the following checklist to help you determine if you are a blackmailer's target.

Do important people in your life:

- Threaten to make your life difficult if you don't do what they want?
- Constantly threaten to end the relationship if you don't do what they want?
- Tell you or imply that they will neglect, hurt themselves or become depressed if you don't do what they want?
- Always want more, no matter how much you give?
- Regularly assume you will give in to them?

- Regularly ignore or discount your feelings and wants?
- Make lavish promises that are contingent on your behavior and then rarely keep them?
- Consistently label you as selfish, bad, greedy, unfeeling or uncaring when you don't give in to them?
- Shower you with approval when you give in to them and take it away when you don't?
- Use money as a weapon to get their way?

If you answered yes to even one of these questions, you are being emotionally blackmailed. But I want to assure you that there are many changes you can put into practice immediately to improve your situation and the way you feel.

COMING TO CLARITY

Before we can affect change, we need to clear away our confusion about how our relationships with blackmailers are operating. We need to turn on the light. This is a vitally important step in ending emotional blackmail because, even as we work to burn off the FOG, the blackmailer is busy pumping in thick new layers. Despite all the sophistication we've developed in recent years about our moods, psyches and motivations, when we're dealing with FOG, our senses are muffled, and the finely tuned sensors that usually guide us in relationships fill with static. Blackmailers can skillfully mask the pressure they're applying to us, and often we experience it in ways that make us question our perception of what's happening. In addition, there is frequently a vast difference between what our blackmailers do and the benign or even loving way they interpret their actions to us and themselves. We feel confused, disoriented and resentful. But we're not alone. Emotional blackmail is a dilemma affecting millions of people.

Throughout this book, in dramatic case histories, you will meet people who are struggling with emotional blackmail—and finding ways to end it. These are stories about real people with real feelings and real conflicts. They're people you'll identify with—men and women who function with great competence, grace and effectiveness in many parts of their lives but who have fallen into the blackmail trap. If you open your heart, you can learn a great deal from them. Their stories are modern fables, teaching stories that can serve as guides and beacons on your own journey.

BLACKMAIL TAKES TWO

In the first half of this book, I'll show you exactly how emotional blackmail operates and why some of us are especially vulnerable to it. I'll explain, in detail, how the blackmail transaction works and show what each party to the transaction wants, what they get and how they get it. I'll explore the psyche of the blackmailer, a task that seems a bit daunting at first because not every blackmailer has the same style or character traits: Some are passive, some quite aggressive. Some are direct and others are extremely subtle. Some let us know precisely what the consequences will be if we displease them, others emphasize how much we are making them suffer. Yet no matter how different they appear to be on the surface, they all have major traits in common, characteristics that feed their manipulative behavior. I'll show you how our blackmailers use FOG and their other tools, and I'll help you understand what drives them.

I'll show how fear—fear of loss, fear of change, fear of rejection, fear of losing power—is a common ground stretching beneath those who become blackmailers. For some, these fears are rooted in a lengthy history of feeling anxious and inadequate. For others, they may be a response to more recent uncertainties and stresses that have undermined their sense of themselves as secure, competent people. I'll demonstrate how the potential for blackmail skyrockets as fears build in the blackmailer's life. And I'll show how triggering events such as rejection by a lover, the loss of a job, divorce, retirement or illness can easily turn someone close to us into a blackmailer.

The people close to us who use emotional blackmail are rarely individuals who wake up every morning saying "How can I destroy my victim?" Rather, they are people for whom blackmail is the ticket to feeling safe and in charge. No matter how confident they look on the outside, blackmailers are operating out of high degrees of anxiety.

But when they snap their fingers and we jump, for a moment our blackmailers can feel powerful. Emotional blackmail becomes their defense against feeling hurt and afraid.

THE ROLE WE PLAY

Without our help, however, blackmail can't take root. Remember: Blackmail takes two—this is a *transaction*—and our next step will be to see what we as blackmail targets contribute.

Each of us brings into any relationship our own potent set of hot buttons—our stored-up resentments, regrets, insecurities, fears, angers. These are our soft spots, places that hurt when touched. Emotional blackmail can only operate when we let people know they've found our hot buttons and that we'll jump when they push them. Throughout this book, we will see how our life experiences have shaped the automatic emotional responses that power our hot buttons.

It's been fascinating to me to watch as our philosophy of human behavior has evolved from one that sees us as victims to one that encourages us to take responsibility for our lives and our problems. Nowhere is that more important than in the arena of emotional blackmail. It's easy to focus on other people's behavior and to think that if *they* change, things will be fine. But what we really need to find is the commitment and courage to understand *ourselves* and to change the way we interact with would-be blackmailers. It's tough to acknowledge that through capitulating, we actually teach the blackmailer how to blackmail us. But the hard truth is this: Our compliance rewards the blackmailer, and every time we reward someone for a particular action, *whether we realize it or not*, we're letting them know in the strongest possible terms that they can do it again.

THE PRICE WE PAY

Emotional blackmail spreads like ivy, and its coiling tendrils can wrap around every aspect of our lives. If we're capitulating to emotional blackmail on the job, we may come home and take it out on the kids. If we have a bad relationship with a parent, we may shower our partner with negativity. We can't just put a conflict in a box labeled *boss* or *husband* and keep it out of the rest of our lives. We may even duplicate the dynamic that's making us suffer and become blackmailers ourselves, taking out our frustration on someone weaker or more vulnerable than we are.

Many of the people who use emotional blackmail are friends, colleagues and family members with whom we have close ties that we want to preserve and strengthen. They may be people we love for the good times we've shared, the closeness we still occasionally manage to have and the histories we have in common. We may consider our relationships with them to be good, for the most part, but pulled off course by blackmail. It's vital not to let the blackmail habit suck us—and everyone around us—into its vortex.

The price we pay when we repeatedly give in to emotional blackmail is enormous. The blackmailer's comments and behavior keep us feeling off-balance, ashamed and guilt-ridden. We know we need to change the situation, and we repeatedly vow that we will, only to find ourselves outwitted or outmaneuvered or ambushed again. We begin to doubt our ability to keep promises to ourselves, and we lose confidence in our own effectiveness. Our sense of self-worth erodes. Perhaps worst of all, every time we capitulate to emotional blackmail, we lose contact with our integrity, the inner compass that helps us determine what our values and behavior should be. Although emotional blackmail is not heavy-duty abuse, don't think for a moment that the stakes aren't high. *When we live with emotional blackmail, it eats away at us and escalates until it puts our most important relationships and our whole sense of self-respect in jeopardy.*

CHANGING UNDERSTANDING INTO ACTION

I have been a therapist for more than 25 years. In that time, I have treated many thousands of people in a variety of settings, and if there is one sweeping generalization I can make without fear of contradiction, it is that *change* is the scariest word in the English language. No one likes it, almost everyone is terrified of it, and most people, including me, will become exquisitely creative to avoid it. Our actions may be making us miserable, but the idea of doing anything differently is worse.

Yet if there's one thing I know with absolute certainty, both personally and professionally, it is this: Nothing will change in our lives until we change our own behavior. Insight won't do it. Understanding why we do the self-defeating things we do won't make us stop doing them. Nagging and pleading with the other person to change won't do it. We have to act. We have to take the first step down a new road.

A NEW VOCABULARY OF CHOICES

All my books have been solution-oriented, and in the second part of this book I will take you step by step through the wide range of choices you have when someone targets you for emotional blackmail. Though we often operate with a constricted view of the choices available to us, we generally have far more options than we realize. And having choice empowers us. I will show you strategies for standing fast in the face of blackmail even when you feel intimidated or afraid, and I'll help you feel good about yourself as you do. I'll provide checklists, simple exercises, practice scenarios and specific nondefensive communication techniques. These are techniques I've taught and refined over the past 25 years—and they work!

Equally important, I will guide you through the vitally important ethical, moral and psychological questions we all struggle with in the face of emotional blackmail, questions like:

- When am I being selfish and when am I being true to my own wants and priorities?
- How much can I do or give without feeling resentful or depressed?
- If I give in to the blackmailer, am I violating my integrity?

I will give you the tools to determine, instance by instance, where your responsibilities to other people begin and end—one of the most important keys to freeing yourself from manipulation.

One of the greatest gifts of this book will be to help you reduce and manage the feelings of guilt that your blackmailers have inspired in you. I will show you how to tolerate the inevitable discomfort that results when you start to change your behavior in the service of breaking free from guilt you don't deserve. I'll show you how guilt diminishes the more you act in healthy, self-affirming ways. And I'll show you that without guilt, the blackmailer is impotent.

I will accompany you through the process of making the major internal shifts that will enable you to abandon your automatic reactions to emotional blackmail and replace them with conscious, positive choices about how far you're willing to go to accommodate someone at the expense of your own well-being.

As I help you resist blackmail, I'll also help you decide when an instance of blackmail is not worth going to the mat about, and even when complying with a blackmailer can actually be smart strategy. In a few extreme cases, the only healthy recourse is to break off from a blackmailer completely, and I'll deal with why and how to do that when all else has failed.

When we finally gain the understanding and behavioral skills that can free us from the deadening cycle of emotional blackmail, we release incredible excitement and energy.

"I was able to say no to my boyfriend and realize that his demands were irrational," my patient Maggie reported to me. "I didn't do anything to hurt him, even though he would like to pretend I did. And for the first time, I didn't go beat myself up and call him back 10 minutes later to apologize or give in."

I wrote this book for everyone who is struggling to stay close to a lover, parent, colleague or friend who's strangling an important and otherwise good relationship with the twisted vine of manipulation.

Please understand that although I can't be right there with you as you travel through this process, I will provide moral support every step of the way as you take these sometimes difficult but life-changing actions. I will help you through the important work of building new and healthy relationships—not only with the blackmailers in your life, but also with yourself.

It takes real courage to confront emotional blackmail. This book will give you the power to do it.

PART I

Understanding the Blackmail Transaction

1

Diagnosis: Emotional Blackmail

The world of emotional blackmail is confusing. While some emotional blackmailers are clear in their threats, others may send us mixed signals, acting kindly much of the time and resorting to blackmail only occasionally. All this makes it difficult to see when a pattern of manipulation is developing in a relationship.

Certainly there are forceful, unambiguous blackmailers who consistently make direct threats about what will happen if you don't go along with them and spell out the consequences of noncompliance in no uncertain terms: "If you leave me you'll never see your kids again." "If you don't support my project, I'll hold up your recommendation until you do." Clear threat, no question of intent.

More often, however, emotional blackmail is far more subtle and occurs in the context of a relationship where much is good and positive. We know what the other person is like at his best, and we let our memories of positive experiences overshadow our nagging feeling that something is wrong. Emotional blackmail creeps up on us, edging quietly over the line from normal, acceptable behavior into transactions that are first tinged, then permeated with elements that compromise our well-being.

Before we can label anyone's behavior emotional blackmail, it has to have certain components. We can make the diagnosis much as a doctor would determine that a person has a physical ailment: by looking at the symptoms. In the example that follows, you'll see a couple in a love relationship, but the symptoms hold true whether the partners in a conflict are friends, business colleagues or family members. The issues may differ, but the tactics and actions will be the same—and clearly recognizable.

THE SIX DEADLY SYMPTOMS

A young couple I know, Jim and Helen, have been together for just over a year. Helen, a professor of literature at a community college, has huge brown eyes and a broad, perfect smile. She was introduced to Jim at a party, and Jim seemed delightful. Tall and soft-spoken, he's a successful songwriter. The couple share a great affection for each other. However, for Helen the ease of being with Jim has been draining away. In fact, their relationship has progressed through the six stages of emotional blackmail.

To give you a clear idea of what the six emotional blackmail symptoms look and feel like, let's walk through a simplified version of a conflict that came up between Jim and Helen. You'll notice that some of the symptoms describe Jim's behavior, and some focus on Helen's.

1: **A demand.** Jim wants something from Helen. He suggests to her that they've been spending so much time with each other, they might as well move in together. "I practically live here already," he tells her. "Let's just make it official." Her apartment is huge, and half his things are there, he adds, so it'll be a simple transition.

Sometimes blackmailers don't verbalize what they want as clearly as Jim did, but instead make us figure it out. Jim might make his point indirectly, perhaps sulking after a friend's wedding, then letting Helen extract from him, "I wish we could be closer; I get so lonely sometimes," and eventually saying that he'd like to move in.

At first blush, Jim's suggestion sounds loving, and not like a demand at all, but it soon becomes clear that he is set in his course of action and is not willing to discuss or change it.

2: **Resistance.** Helen feels uncomfortable about Jim's moving in and expresses her unwillingness by telling him that she's not ready for that kind of change in the relationship. She cares deeply for him, but she wants him to have his own place.

If she were a less direct person, Helen might resist in other ways. She could withdraw and become less affectionate, or tell him that she's decided to repaint and he'll have to take his things home until the job is done. However she expresses her resistance, the message is clear. The answer is no.

3: **Pressure.** When he sees that Helen isn't giving him the response he wants, Jim does not try to understand her feelings. Rather, he pushes her to change her mind. At first he acts as if he's willing to talk over the issue with her, but the discussion becomes one-sided and turns into a lecture. He transforms Helen's statement of resistance into a statement of her deficiencies, and he casts his own desires and demands in the most positive terms: "I only want what's best for us. I only want to give to you. When two people love each other, they should want to share their lives. Why don't you want to share yourself with me? If you weren't so self-centered, you could open up your life a little."

Then he turns on the charm and asks, "Don't you love me enough to want me here all the time?" Another blackmailer might turn up the pressure by adamantly insisting that his moving in would improve the relationship and bring them closer. Whatever the style, pressure will come into play, though it may be cloaked in benevolent terms—for example, Jim's letting Helen know how much her reluctance is paining him.

4: **Threats.** As Jim continues to hit a wall of resistance, he lets Helen know there will be consequences if she doesn't give him what he wants. Blackmailers may threaten to cause us pain or unhappiness. They may let us know how much we're making them suffer. They may also try to tantalize us with promises of what they'll give us, or how much they'll love us, if we go along with them. Jim works on Helen with veiled threats: "If you can't make this kind of commitment to me after all we've meant to each other, maybe it's time for us to see other people." He doesn't directly threaten to end the relationship, but the implication is impossible for Helen to miss.

5: **Compliance.** Helen doesn't want to lose Jim, and tells herself that perhaps she's been wrong to say he can't move in, despite her continuing uneasy feelings. She and Jim only talk superficially about her concerns, and Jim makes no attempt to allay them. A couple of months later, Helen stops resisting, and Jim moves in.

6: **Repetition.** Jim's victory ushers in a quiescent period. Now that he's gotten his way, he removes the pressure, and the relationship appears to stabilize. Helen is still uncomfortable about how things have turned out, but she's also relieved to have the pressure off and to regain Jim's love and approval. Jim has seen that pressuring Helen and making her feel guilty is a sure way to get what he wants. And Helen has seen that the fastest way to end Jim's pressure tactics is to give in. The groundwork is laid for a pattern of demands, pressure and capitulation.

These six characteristics are at the heart of the emotional blackmail syndrome, and we will be returning to them and exploring them more deeply throughout this book.

IF IT'S SO CLEAR, WHY DON'T WE SEE IT?

These symptoms seem so clear, and so troubling, that you'd think alarm bells would go off when they begin to appear. But often we're immersed in emotional blackmail before we recognize that it's entangled us. In part this happens because emotional blackmail takes to extremes behavior that we use and encounter all the time: manipulation.

Many forms of manipulation aren't troublesome at all. We

all manipulate one another at times, and we all get manipulated. We've learned to play a multitude of games to maneuver people into doing what we'd like. One of my favorites is "Gee, I wish someone would open the window," as opposed to "Could you please open the window?"

It's amazing how hard it is for many of us to be direct even about minor things, much less when there's a lot at stake and we want something significant. Why not just ask? Because asking is risky. What if the other person says no? Letting people know what we want in a direct and clear way is something few of us do. We're afraid to put ourselves on the line by telling the other person what we want or how we're feeling. What if we end up feeling angry—or worse yet, rejected? If we don't actually ask, then if the other person says no, it's not really a no, right? We can explain away any discomfort.

We can also avoid appearing too aggressive or too needy if we don't make an outright request. It's easier to find indirect ways to signal people in hopes that they'll read between the lines and figure out what we want: "The dog looks like he needs to go out [hint, hint]."

Sometimes we even do this without words. Obvious or subtle hints—a sigh, a pout, what's known as "that look," we all use them, and we're all on the receiving end, even in the best of relationships. But there is a clear point where everyday manipulation turns into something far more harmful. *Manipulation becomes emotional blackmail when it is used repeatedly to coerce us into complying with the blackmailer's demands, at the expense of our own wishes and well-being.*

THE RIGHT TO SET LIMITS

When we're talking about emotional blackmail, we're automatically talking about conflict, power and rights. When one person wants something and the other doesn't, how hard can each reasonably push? When does another person's pressure go too far? This is fuzzy territory since we now put so much emphasis on expressing feelings and setting limits. Remember, it's important that we not label every conflict or expression of strong feelings or especially instances of healthy limit-setting as emotional blackmail.

To help you discern the distinctions more clearly, I'd like to show you several situations that involve appropriate limit-setting, then let you see what they look like when they've crossed the line into emotional blackmail.

No Blackmail Here

Not long after she sold a photography book she'd been working on for almost a year, my friend Denise related a situation that had come up between her and Amy, a friend who'd been her colleague at an ad agency until they'd both decided to freelance. Denise wondered if Amy was emotionally blackmailing her.

As Denise told me,

From the beginning, we'd always been able to talk about anything. We've spent hours comparing notes about the struggles of being on our own and the challenges of life after downsizing—both of us started out working for big companies, and both of us sometimes miss it. We talk a lot about the scariness of being on our own, and we do lots of supportive things for each other. We were really close until I told her about working on this book.

She sounded happy for me, but soon after, she called me up and told me, "You know, I'm feeling kind of jealous. I'm working so hard right now and not much is happening. I'd appreciate it if you just didn't talk about your work and how excited you are for awhile—it's a sore spot with me." So—I said OK. And as though nothing had happened, we switched gears and started talking about what she's working on.

Now, if I even mention something about the book, she'll stop the conversation and say, "It would be best if you and I didn't talk about that." It's beginning to feel like a strain, but I like her, and I'm trying to adjust to playing by her rules.

At first blush it may appear that Amy is pressuring Denise in order to get her own way and controlling the interaction between them by deciding what is and isn't OK to talk about.

But in reality, Amy is honestly acknowledging her feelings and taking good care of herself by setting limits on how much she can hear of Denise's good news. Amy has the right to do that. It's only human to be envious when someone gets something we want for ourselves, especially if we're at a low spot in our lives. There are times when all of us want to avoid certain topics, and like Amy, we have a right to set limits. Denise also has the right to decide that she doesn't like the limits Amy has set and to express *her* discomfort, or to spend less time with Amy.

In this situation, Amy has made no threat, either direct or indirect, about what she will do if Denise doesn't honor her request. There's also no real pressure, just a statement of needs and feelings. Yes, there's a conflict. Yes, Denise is uncomfortable with the change in the relationship. Yes, there are some powerful feelings here. But no, this isn't emotional blackmail.

Crossing the Line

Now let's look at the same situation and *add* the component of emotional blackmail. I think you'll see how the atmosphere and the scenario change. Say that on hearing Denise's news, Amy's response is something like this: "I'm so glad to hear about your book project! I know you'll have a ton of work. Wouldn't it be great if we could work together on it? I could be your assistant."

When Denise says she doesn't need that kind of help, Amy says, "I thought you were my friend. You know things are rough for me right now. It was tough enough breaking up with Roger, and on top of that, you know how bad the money's been since I got that huge tax bill. I've been so depressed I can hardly work. I thought you were the kind of person who'd help a friend in need."

As she meets continuing resistance, Amy turns up the pressure, appealing to Denise's generosity. "I can't see how it would hurt you to share your good fortune with me," she says. "You know I would do the same for you." She begins to label Denise selfish and greedy, and emphasizes the bleakness of her own situation. At the same time, she threatens to cut off the friendship if she can't become Denise's assistant. Finally, Denise gives in.

This scenario contains all the elements of emotional blackmail: a demand, resistance, pressure, threats and compliance. And it's a scenario just ripe to be repeated.

ONE CONFLICT, TWO SOLUTIONS

Asking someone to refrain from talking about a touchy subject is fairly innocuous. But what if the conflict involves something more serious: a partner's affair, another person's drinking problem, someone's dishonesty at work? In such cases, people may say terrible things to each other, and setting limits may begin to sound like emotional blackmail because feelings are so strong. Yet even here, there's a sharp distinction between appropriate limit-setting and emotional blackmail. Again, let's contrast a couple of similar situations.

The Affair

I've known my friend Jack and his wife, Michelle, for years, and I've always admired their marriage. There's a large age difference between the two—Jack is 15 years older—but the two of them, both musicians in the symphony, seem to have a rare kind of intimacy. One night, Jack offered to drive me to a meeting of the opera group we both belong to, and on the way home we had a chance to talk. "What makes it work between you two?" I asked. "Who gave you the secret of a perfect marriage?"

Jack's answer wasn't what I'd expected.

To tell you the truth, things haven't always been so perfect. At least I *haven't been. I'm going to tell you something very few people know. Three years ago, I did a stupid thing. I started seeing a young woman who was playing with the orchestra as a guest violinist. The affair was short, but I felt guilty as hell. It was dumb. Thoughtless. I couldn't stand keeping that secret from Michelle, and I knew that I could never be genuinely close to her again unless I confessed. So I made the decision that it was best for me to tell her and risk whatever followed.*

At first I thought she was going to kill me. She barely spoke to me for a couple of weeks, and I moved downstairs into the den. But then she surprised me. She said she'd been thinking, and she realized that we needed a plan if we were going to be together for the rest of our lives. She said she was mad as hell, but she wanted to offer me a deal: She'd let the subject drop and she wouldn't keep rubbing my nose in what I'd done or use it as a club when she wanted something from me. But if I didn't recommit to her in an exclusive relationship and knock off the crap and go to counseling with her, there was no way we were going to make it together. And if I couldn't make those commitments, she wouldn't be able to stay in the marriage, because she wasn't going to live with insecurity and uncertainty and suspicion.

I told Jack he was fortunate to be dealing with Michelle because she set limits in a healthy way, a process I'll outline here and discuss in greater depth in the second half of the book. In dealing with Jack, Michelle:

- Defined her position
- Stated what she needed
- Laid out what she would and wouldn't accept
- Gave Jack the choice of saying yes or no

Also, she insisted on therapy for *both* of them.

All of us have the right to let others know when their actions are unacceptable to us, just as Michelle did. We all have a basic right not to live with poison in our relationships, whether it's dishonesty or addictions or any form of abuse.

If someone confronts us fairly about something we've done, the words and feelings may be strong, but if there are no threats and no pressure, there is no blackmail. Appropriate limit-setting isn't about coercion, pressure or repeatedly characterizing the other person as flawed. It's a statement of what kind of behavior we will and won't allow into our lives.

A Blackmailer's Approach

Contrast Michelle's way of dealing with this crisis with that of a couple I saw as clients a couple of years ago. Stephanie and Bob's marriage was on the brink, and they came into my office barely speaking to one another. An attractive couple in their late 30s, Bob is a tax attorney with an active practice, and

Stephanie sells real estate. Since it was Bob's idea that they come to see me, I asked him to start the session.

He began,

I don't know how much more I can take. I made a horrible mistake 18 months ago and it's destroying us. I had a short affair with a woman I met on a business trip. I'm totally to blame. This never should've happened. But it did. And I'm doing my best to find ways to make it up to Stephanie because I love her and I want to stay with her. We have a good life, two beautiful kids. But my God I'm being treated like a mass murderer. She just won't let go of it.

Now she brings it up every time she wants something. She dictates when her parents will come and stay with us, something as silly as which movie we're going to see, what I have to buy her to make her happy. Right now she wants to go to Europe at a time when I have a big case coming up, and there's no way I can take off. I'd love it if she went and took a friend, but she wants what she wants when she wants it. I'm supposed to drop everything and go right then. It's like I have to because I betrayed her. She'll say, "You owe me. If you live to a thousand you can't make up what you did to me." If I don't give her her way all the time, she's going to remind me of what a rat I was. She even put a Post-it on the medicine cabinet saying "cheater." How can I not give in to her? I'm afraid she'll leave me if I don't. It's true. I was a cheater and I feel awful about it. But I can't go on like this. How do we get out of this swamp?

Stephanie, like Michelle, had a right to be angry. But her response to Bob was punitive and controlling. In fact, it was blackmail. Frightened and insecure after she learned of Bob's affair, Stephanie erroneously believed that she could bind him to her by making him feel so guilty that he would do whatever she wanted. She continually defined him as morally inferior and unworthy, using his transgression as a weapon. Her threat was clear and constant: "If I don't get what I want, I will make you miserable." Her message: "I am in control now."

A crisis such as an affair can be an experience full of both danger and opportunity. It's also one of those complex life situations that's rife with potential for blackmail. Michelle used her experience as a chance to refocus her relationship with Jack and to define what she expected of him, herself and their marriage. But Stephanie got mired in rage and revenge.

The possibilities for hurt or healing exist in any situation in which we've chosen to maintain a relationship after a serious transgression: a betrayal by a colleague, a damaging rift in a family, a discovery that we've been deceived by a friend. But if both parties are coming from a position of goodwill and truly want to resolve whatever crisis is impairing the relationship, there is no place for emotional blackmail.

WHAT'S THE REAL MOTIVE?

How can we know if someone is more interested in winning or in resolving the problem? They're not going to tell us. They're certainly not going to come out and say, "I don't care what you want, I'm only trying to get my way." In an emotionally intense situation, our perceptions get clouded, a condition that gets worse when we're feeling pressured. The following list will help you diagnose emotional blackmail by allowing you to clarify the intentions and goals behind the other person's behavior.

If people genuinely want to resolve a conflict with you in a fair and caring way, they will:

- Talk openly about the conflict with you
- Find out about your feelings and concerns
- Find out why you are resisting what they want
- Accept responsibility for their part of the conflict

As we saw with Michelle and Jack, you can be angry with someone without trying to beat them up emotionally. Disagreements, even strong ones, don't have to be mixed with insults or negative judgments.

If someone's primary goal is to win, he or she will:

- Try to control you
- Ignore your protests

- Insist that his or her character and motives are superior to yours
- Avoid taking any responsibility for the problems between you

When you see that other people are trying to get their way regardless of the cost to you, you're looking at the bottom-line behavior of the emotional blackmailer.

FROM FLEXIBLE TO FROZEN

In looking at situations that might be drifting into emotional blackmail, as well as probing for symptoms and motives, I ask another question: How much flexibility do I have, and allow, in this relationship?

When emotional blackmail begins to seep into a relationship, we feel a major shift in the atmosphere that surrounds us. As we saw with Stephanie and Bob, the relationship gets stuck. Threats and pressure become a regular part of our everyday interactions. A chill sets in, and we lose much of the flexibility that allows us to navigate smoothly through the rough spots.

When flexibility is there, we can easily take it for granted. Every day, without much trouble or trauma, we negotiate the myriad details of our existences—what restaurant to eat in, what movie to go to, what color to paint the living room or where to hold the office picnic. Actually, in many cases, the outcome doesn't matter too much, and the person with the strongest preference usually wins by default. But despite normal disagreements and manipulation, there's a rhythm of give and take, a sense of balance and fairness. We can give in to many things with little negative effect and replenish our egos and energies quickly. At the same time, we also expect others to give us *our* way from time to time.

When the willingness to compromise begins to disappear, the status quo becomes the template for the future. It's as though we're not allowed to change, or move away from a role that may not always fit. We're frozen.

When I was a kid, my friends and I used to play a game called freeze tag in which the object was to escape being touched by the person who was It. If you were tagged, you had to freeze in place and not move until the game was over. A lawn where the game was in progress wound up looking like a sculpture garden, scattered with kids holding themselves in odd, shocked-looking positions. Emotional blackmail is a lot like freeze tag, but it's not a game anymore. Once blackmail has touched a relationship, it becomes rigid, stuck in patterns of demands and capitulation. We're not allowed to adjust our stances or change positions.

Allen is a bright, funny man who owns a small furniture company. But he was grim the first time he came to see me as he described his problems with his new wife, Jo.

"I thought Jo was just what I wanted—she's gorgeous, she's got a great sense of humor, she's smart," he began.

"Sounds good to me," I said. "So why the long face?"

I just don't know if this is going to work. I know she loves me, but I don't like what's happening to us. If I suggest that we spend any time apart—my friends are always bugging me to go to movies or hang out after work—she acts really hurt. She looks at me with those big sad eyes and says, "What's the matter? You're bored with me? You don't want to be with me anymore? I thought you were wild about me." If I start to make plans, she pouts, pleads with me and lets me know how unhappy I'm making her in no uncertain terms. I didn't know she was so needy. I'm fine if she wants to go off with her friends, but she wants to do less and less of that. It's like she wants to be in my pocket. One time I actually got brave enough to go out with my friends anyway, and she wouldn't speak to me for the rest of the week. I thought she was the one—she's great—but I'm feeling a lot of resentment. We have such a good relationship in so many ways. But dammit, she sure likes to get her own way.

Often when needy, dependent people get into relationships, they panic when their partner wants to enjoy activities apart from them. They experience abandonment fears and rejection anxiety, and rather than talking about those fears, they hide them. They're adults, after all, and "supposed" to be independent, not feeling like scared little kids. As Jo saw Allen wanting more freedom, instead of being able to talk about her feelings, she expressed them indirectly. She made Allen feel guilty when he wanted to do a perfectly normal thing like going out by himself.

Allen was doing his best to understand her.

She had a tough time as a kid, so I know why she's so needy. I don't blame her for feeling insecure. Some days it feels great to have a woman who wants me so much she won't let me out of her sight. But to tell you the truth, it's starting to get me down. She gets what she wants by making me feel so damn guilty all the time. And I feel like a wimp always giving in to her.

Though he didn't want to admit it to himself, Allen realized that behind Jo's pleading looks and those charming, loving-sounding statements, there was a demand, backed up by well-placed pressure. Jo expected him to spend all his free time with her—that was the only role she willingly allowed him to take. He wasn't allowed to have activities or interests of his own. But Allen did something that many targets of emotional blackmail do, particularly at the beginning: He gave Jo the benefit of the doubt and rationalized away her clinginess because of his compassion for her difficult childhood and his deep feelings for her now.

He also did what many people do when they are pressured by neediness or possessiveness: He misinterpreted it as a sign of how much she cared about him. As we will see throughout this book, understanding and compassion get you nowhere with an emotional blackmailer. In fact, they only add fuel to the blackmail flames.

Once you discern the symptoms of emotional blackmail in any relationship, you may feel as though a rug's been pulled out from under you. Suddenly you realize you don't really know your lover, or your parent, or your brother, or your boss, or your friend. Something's been lost. There's little room for compromise or flexibility. There's no balance of power, no sense that one of you gets your way one time, the other the next. Where no "payment" was previously required for love and respect, being in the good graces of blackmailers increasingly depends on giving them their way.

2

The Four Faces of Blackmail

"If you really loved me . . ."
"Don't leave me or I'll . . ."
"You're the only one who can help me . . ."
"I could make things easy for you if you'd just . . ."

In blackmail-speak, all of these statements are ways of putting a demand on the table. Yet they appear to be markedly different because each reflects a singular blackmail type. When we look closely at emotional blackmail, what seems to be one kind of behavior splits into four varieties, distinct as the bands of color that appear when you pass a beam of light through a prism.

Punishers, who let us know exactly what they want—and the consequences we'll face if we don't give it to them—are the most glaring. They may express themselves aggressively, or they may smoulder in silence, but either way, the anger they feel when thwarted is always aimed directly at us. **Self-punishers,** who occupy the second category, turn the threats inward, emphasizing what they'll do to *themselves* if they don't get their way. **Sufferers** are talented blamers and guilt-peddlers who often make us figure out what they want, and always conclude that it is up to *us* to ensure that they get it. **Tantalizers**

put us through a series of tests and hold out a promise of something wonderful if we'll just give them their way.

Each type of blackmailer operates with a different vocabulary, and each gives a peculiar spin to the demands, pressure, threats and negative judgments that go into blackmail. These differences can make blackmail difficult to detect, even when you think you're savvy enough to recognize it. If you think all birds look like eagles, you might be shocked when someone tells you the swan that just floated by was a bird, too. The same kind of cognitive dissonance occurs when an unexpected form of emotional blackmail appears in your life.

But once you understand the four faces of blackmail, you can begin to see the danger signs in another person's actions and develop an early-warning system that can help you to predict, prepare for and even prevent emotional blackmail.

THE PUNISHER

I've begun this field guide to the blackmail quartet with the most blatant blackmailers—the punishers—not because they're necessarily the most prevalent but because they're the most obvious. It's impossible to be unaware that you've crossed punishers, because any perceived resistance incurs immediate anger. They may express that anger aggressively, with direct threats—I call this group the active punishers—or unleash a smouldering fury, the mode of passive punishers. Whatever their style, punishers want relationships in which the balance of power is totally one-sided: "My way or the highway" is the punisher's motto. No matter what you feel or need, punishers override you. They cancel you out.

Active Punishers

"If you go back to work, I'll leave you."

"If you don't take over the family business, I'll cut you out of my will."

"If you try to divorce me, you'll never see your kids again."

"If you won't accept this overtime, you can forget about asking for a promotion."

These statements are strong—and they're scary. They're also highly effective because they give us a sharp picture of what will happen if we don't give aggressive punishers their way. They can make our lives miserable, or at the very least unpleasant. Punishers may not always realize the full impact of their words or notice how often they threaten to disapprove of us, let others know how terrible we are or take away something that is important to us. Punishers may not carry through on a threat 19 times out of 20, and they may be quite pleasant in the calms between storms, but because the threatened consequences can be serious, we live in fear of the one time the active punisher follows through.

Liz, a thin, dark-eyed woman with a low, quiet voice, came to see me, as so many women have over the years, for help in figuring out if there was anything she could salvage of a romance that had become a cold and emotionally abusive marriage. She had met Michael at a training session for computer salespeople several years out of high school, and as they worked on a team project, she'd been impressed by his authoritative way with people and his ability to cut to the core of any problem. His good looks only sweetened the package.

Michael seemed terrific in the beginning. He was caring and responsible, and there are still a lot of good times. So it took me a long time to see what a control freak he was. Within a year after we were married, I was pregnant with the twins, and I settled right into the routines and demands of motherhood. When the twins were finally in school, I thought I'd better head back to school myself. In our field, you have to either keep up or get out. But Michael believes that mothers of school-age kids belong at home. Period. He's swatted me down every time I've asked for his ideas about what we can do about day care and tuition.

I got so frustrated that I told him I didn't know if I wanted to stay in a marriage with him. And that's when things got crazy. He told me that if I leave him, he'll get all the money and leave me out on the street. He was like a total stranger. "You like living in a nice house? You like your lifestyle?" he said to me. "You just think about filing for divorce and I'll see you out on the curb like a bag lady. And once my lawyer gets through with you, they'll never let you near the kids again. So knock off the crap about divorce and behave yourself." I have no way of knowing if he's full of hot air or if he'd go that far to hurt me. So I told my lawyer to stop everything and I canceled the divorce proceedings. Right now I just hate him, and I don't know what to do.

As Liz found, there is no more fertile ground for punishers than marital difficulty, the end of a love relationship or a divorce. Perhaps the most powerful blackmailers are individuals like Michael, who at times of overwhelming stress and pain can threaten to make their target's life even more miserable by cutting off financial resources or contact with the children—and who pour on any other punishment they can think of.

People trying to deal with punishers are always between a rock and a hard place. If they resist and try to hold their ground, they run the risk that the punisher will actually carry through on the threat. If they capitulate, or at least buy some time, they find themselves swimming in a cauldron of rage: rage at the blackmailer for creating such an oppressive and constricting situation, and at themselves for not having the guts to put up a fight.

Kids R Us

It's not surprising that many of the people who've mastered the art of turning otherwise competent adults into children are parents. Often our parents need to maintain control over us long after we've left the nest, feeling it is their job to help determine whom we should marry, how we should raise our children, where we should live and how. They can wield enormous power because of our loyalty to them as well as our lifelong fear of incurring their disapproval. That fear can loom large when parents turn up the pressure on us, using wills or the promise of money to cement their authority and our obedience.

My client Josh, a 32-year-old furniture designer, has met the love of his life, an enthusiastic businesswoman named Beth. He's wildly happy, and there's only one problem: his father, Paul.

My dad has always been devout—we're Catholic—and everyone has always married within the church. Just my luck, I fell in love with a Jewish girl I met playing racquetball! I tried to talk to Dad about it, but he went crazy. If I marry Beth, he's threatened not to co-sign the loan for my business, which I've built all my plans on, and to cut me out of his will. And you know what? He might do it. I can't bring Beth home or even mention her, and it's ridiculous. There's no point in talking to Dad about this—I've tried. He makes the pronouncement that this subject is not to be discussed again and walks out of the room. I keep asking myself, Am I for sale? How much is my soul worth? Am I supposed to stop seeing my family, or is it better to go on lying to them and pretending Beth doesn't exist? This is killing me. It's more than money—I've always been close to my family, and now I can't go home without lying.

Parents who are punishers often make their children choose between them and other people they love, setting up a situation in which any choice is seen as betrayal. And their blackmailed children often cling to the fantasy that if they give up a "flawed" partner for the sake of making peace in the family this time, next time they'll be able to find someone who passes the test. Of course, it *is* just a fantasy. Parents using this kind of blackmail will inevitably find a flaw in the next person, and the next, and in anyone who represents a threat to their control.

Josh bent over backward trying to placate Paul while at the same time holding on to what he wanted for himself. No matter which way he looked, he could see only choices that compromised his integrity. He could give in to his father, which wasn't an acceptable option—he had no intention of giving up Beth—or he could pretend to go along and live a lie.

As we maneuver to avoid the wrath of punishers and the aggressive way they manipulate us, we may find ourselves doing things that amaze us—lying, keeping secrets, sneaking around—to maintain the illusion of obeying them. Behaving as though we're still rebellious teenagers, and violating our own standards, adds to what may already be a significant load of self-reproach born out of the failure to stand up to the blackmailer.

The Silent Treatment

Punishers needn't be articulate, or even speak, to get their message across. As compelling as the blackmailers we've seen so far are the sulkers and nontalkers who retreat behind unverbalized anger.

Jim, the songwriter we met in the last chapter, revealed himself as a silent punisher soon after he moved in with Helen, and her description of his style of pressuring her is a prime example of this style of blackmail.

I don't know what to do about Jim. When he gets upset with me, he clams up and goes a million miles away. I know he's mad, but he won't talk it out. The other night, I got home late with a pounding headache. My classes had been grueling, and my department head needed a staffing report from me to submit with his budget—no rest for the weary. Jim made me dinner, lit some candles and gave me a fabulous welcome home, and I was really touched. He's such a sweet man. When he wanted to cuddle on the couch later on, I knew what was coming—he wanted to make love. Ordinarily, that would be a treat. But my head was still pounding, I was still thinking about things I hadn't done at work, and I felt about as sexy as a paper bag. I tried to tell him as nicely as I could that this was not a good time for me, and could I have a rain check. But he took it all wrong. He didn't yell, he didn't even say a word. He tightened his jaw, gave me one of those dark looks of his and walked away. Next thing I knew, the den door slammed shut and the stereo was blaring from inside.

The hard, cold silence of these punishers is difficult for almost anyone to bear, and we'd almost sell our souls to avoid having to live with it. "Say anything," we plead. "Yell at me. Anything is better than this silence." Usually, the more we try to get sulkers to tell us what's wrong, the more they withdraw and withhold, terrified of confronting us or their own anger.

I didn't know what to do. I felt this horrible pang of guilt. He'd been so romantic, and I was so cold. So I went in and tried talking to him. He sat there and looked right through me, then he said, "Don't talk to me." I had to do something to make up. So I put on a white satin nightgown, went back into the den, put my arms around his neck and told him how sorry I was. We ended up making love right there. It sounds kind of sexy, but it wasn't at all for me. I still had this headache, I felt so tense I could've snapped in two, and it was awful. But I guess I was desperate to get him to talk to me. I couldn't stand the silent treatment one more time.

Silent punishers barricade themselves behind an impenetrable facade and deflect any responsibility for their feelings onto us. Like Helen, we're in turmoil when someone is punishing us this way. We can feel their anger silently building, and we know that we are its target. They trap us in a pressure cooker of stress and tension, and most of us, like Helen, will quickly give in to them because it's the fastest way we know to find relief.

Double Punishment

When you have a dual relationship—your lover is your boss, or your best friend is your business partner—the potential for punishment rises exponentially. A punisher can, and often will, carry any turmoil from one relationship into the other.

My client Sherry is 28, an ambitious, model-pretty woman who was extremely agitated when she first came to me, trying to break off a romance with the man she works for. Sherry decided to become a secretary to get an inside look at the film business and quickly became assistant to the head of a special-effects house, a volatile 52-year-old filmmaker named Charles. Charles, like Sherry, had an Ivy League education, and they share a love for obscure silent movies and modern art. Sherry was immediately attracted to a man who would take her seriously. Their conversations are lively, and because of her position Charles makes her privy to the inner workings of his business and deal-making. He's been grooming her for months to become his operations manager, a position from which she can participate in meetings with clients and help shape the business.

Sherry's friends warned her not to get involved with her boss, particularly since he's married, but Charles was more interesting to her than men her age, and though she wasn't initially attracted to him, the long hours and the intensity of their work together pulled them close. The building sexual tension between them finally culminated in an intense romantic relationship.

I know, I know. Rule number one is never, ever have an affair with your boss. But Charles is a remarkable man. No one ever got to me in the way he does. I loved seeing how his mind works, and also his worldliness. He has so much to teach, and I felt like I won the prize of being his star student. I loved the intimacy, knowing how much we share. I know we have the same vision for the company. He can't talk to his wife about his work—she's an alcoholic, and she's floating off in space somewhere. Even before we got romantically involved, he was saying that once she got stable enough to stand on her own two feet, he was going to leave her. So I jumped in.

The romance was heady, the sex satisfying, the job experience rewarding. But two years passed, and Charles made no move to separate from his wife. As time went on, Sherry could see that there was no end in sight for his marriage.

After two years of broken promises it finally sunk in that Charles was perfectly content to have a wife and *a mistress, and I didn't like seeing myself permanently in that role. I want to have a real family someday. We were having dinner, and he started telling me about the Paris vacation he was planning to take with his wife and daughter. He knew how much I love Paris, and how we'd talked about getting married there, and I knew then that I'd been living in Fantasy Land. I was a wreck trying to deal with that realization, and finally I told Charles that I wanted to move our relationship back to that close, nonsexual place where we'd started. It would be sad, but we'd both be able to get on with real life.*

Charles had been so generous and kind to me that I was

shocked by how out-of-character his response was. He told me that if I stopped seeing him, I could kiss him—and my job—good-bye. I don't know if I can deal with this breakup and unemployment at the same time. I'm finally close to a job I love, and I'm afraid he's going to slam the door in my face. But staying with him would be like prostitution. I couldn't look myself in the eye. I can't believe I'm even considering it.

Charles was facing loss of a passionate relationship that probably made him feel young and alive, and when he saw that he might lose it, he became desperate and lashed out in the hope of hanging on. That response was a shock to Sherry, but in the white heat of the end of a love affair, it wasn't at all atypical.

Sherry was left to face something that people, especially women, have had to deal with for years. It's always perilous to become intimately involved with a person who's in a position of power over you. If a rift opens in the relationship, you may discover, as Sherry did, that the stress and disappointment can trigger a punitive response from someone who has been an integral part of the fabric of your life. But as we'll see later, Sherry wasn't backed into a corner as much as she thought. There were options available to her, which we will explore later in this book.

Their Blind Spots—and Ours

The closer the relationship, the higher the stakes—and the more vulnerable we are to punishers. We don't want to walk away from people we care deeply about and with whom we have longtime, or even lifelong, bonds—or, in instances like Sherry's, people we depend on for our paychecks as well. We'll often go to great lengths to avoid standing up to such people. We allow ourselves to believe our punishers and to let our most useful insights about their actions fall into a black hole. Josh, for example, simply couldn't see that while his father insisted he had Josh's best interests in mind, his demands were totally self-centered and had little to do with Josh's feelings. The demands of punishers rarely do.

When blackmail escalates, the threatened consequences of not acceding to a punisher can be alarming: Abandonment. Emotional cutoff. Withdrawal of money or other resources. Explosive anger directed at us. And, at the most terrifying extreme, threats of physical harm. The darkest threats, of course, turn into emotional abuse as they escalate into intimidation, and one person seizes total control.

Obviously, in the heat of emotional blackmail, blinded by the intensity of their own needs, punishers seem to be oblivious to our feelings and not terribly introspective about their own behavior. They genuinely believe in the correctness of what they're doing and the rightness of what they want. Confronting punishers can take tremendous inner resources, but it can be done. With tools and guidance, all of the punishers' targets we've met were able to reclaim their adult confidence and were finally able to say—and to show—that they wouldn't be blackmailed anymore.

THE SELF-PUNISHER

We've all met the little six-year-old terror who signals a tantrum by announcing loudly, "If you don't let me stay up and watch videos I'll hold my breath until I turn blue!" Adult self-punishers are a bit more sophisticated, but the principle is the same. They inform us that if we don't do what they want us to do, they'll be upset and maybe even unable to function. They may vow to do something to screw up their lives, or even hurt themselves, because they know that the way they can manipulate us the most successfully is to threaten to undermine their own health or happiness. "Don't argue with me or I'll get sick or depressed." "Make me happy or I'll quit my job." "If you don't do X, I'll stop eating, stop sleeping, drink or take drugs. I'll ruin my life." "If you leave me, I'll kill myself." These are the self-punisher's threats.

Allen, the businessman we met in Chapter 1, slowly began to realize that his new wife, Jo, was blackmailing him by describing what would befall her if she didn't get her way. Over time, her incessant demands on his time and her unwillingness to seek out independent activities had begun to feel increasingly oppressive to him.

I'm not sure I'm ready to do anything drastic, but nothing seems to get through to her. I've tried to talk to her about the fact that things aren't working, but she refuses to have that conversation. She gets quiet, and sometimes I'll see her eyes start to tear up a little. Then she goes into the bedroom and locks the door. I plead with her to come out, and eventually she starts to talk—or should I say turn on the blame machine.

This last time, all I wanted to do was go up to my sister's cabin in Oregon and hang out with her. You'd think I was leaving the planet with no forwarding address. "You know I won't sleep when you're not here and I won't be able to work," she told me. "I need to have you with me. And this is a really stressful time for me. You know I count on you to help me gear up for the big sales season. If you're not here to keep me organized, everything will fall apart. I couldn't possibly do everything I need to under that kind of pressure. Don't you care that I need you? Is that what you want, to screw up my whole life so you can go off for a week?"

I said to her, "For chrissake, it's not the end of the world. I just want to spend some time with my sister." But in her mind, I was abandoning her. I canceled the trip. I'm pretending I never wanted to go. It might not be so bad. She's been so sweet and affectionate since I said I'd stay home that it feels like another honeymoon. But I have these moments when I feel like I just can't breathe.

High drama, hysteria and an air of crisis (precipitated by *you*—of course) surround self-punishers, who are often excessively needy and dependent. They tend to fuse and enmesh themselves with those around them, and often have a real struggle taking responsibility for their own lives. If they turn to blackmail, they justify their demands by making every difficulty, real or imagined, your fault. In fact, they have an incredible talent for making you feel responsible for what happens to them. Where punishers turn their targets into children, self-punishers cast their targets in the role of the grown-up—the *only* adult in the relationship. We're the ones who are supposed to come running when they cry, comfort them when they're upset, figure out what's making them uncomfortable and fix it. We're the competent ones who can save them from themselves, rescue them from their helplessness, protect their fragile beings.

You'll Ruin My Recovery

One of the calls I got most frequently on my radio program was from middle-aged parents desperate to know how to deal with an adult child who was abusing drugs, refusing to work or go to school and draining the family's resources. Whenever the parents tried to change the situation, the threats came thick and fast. "OK. I'll leave. And I'll bet you'll enjoy seeing me out on the street. You never loved me anyhow." "I'll just become a prostitute and then you'll be happy." The intimidated parent would agree to maintain the status quo, even though it was destructive for everyone.

My client Karen, a retired nurse in her late 50s, is working hard in therapy on her relationship with her daughter Melanie. To help Melanie overcome a serious drug dependency, Karen paid for an expensive rehab program, sought counseling for herself as well as joining Al-Anon and encouraged Melanie to enter a training program at the hospital where she used to work. Karen didn't expect gratitude, but she didn't expect blackmail, either.

Melanie is a great kid, and I'm proud of what she's done to turn her life around. But we fight all the time about money. When she got married to Pete, they wanted to buy a house, and they asked me to loan them the money for a down payment. You know what a nurse's pension looks like. I would've loved to help them out, but I didn't have savings to draw on unless I emptied out my retirement account, and I was scared to do that. It's all I have for the future. But the message from Mel was, Why should I have money and she have to do without? She had to have that house.

I worry because I think her sobriety is still a little shaky, and she's not that strong yet. So it's kind of, you know, If you're not careful with me, I'm liable to go back to using and drink-

ing. The threat is, If you don't treat me the way I want to be treated, I'll start drinking again. I have no choice—I have to help pay for that house.

Karen's statement that she had no choice is one I hear often from targets of blackmail, and it reflects the sense of victimization that targets feel. Karen actually had several choices, but it would take some work for her to be able to see and use them. Melanie's threat to go off the wagon went straight for Karen's jugular. As I pointed out to Karen, it was a strong-arm tactic that didn't mesh at all with her description of Melanie as weak, the label that many self-punishers use for camouflage.

The Ultimate Self-Punishment

The ultimate threat self-punishers can make is frightening in the extreme: It's a suggestion that they will kill themselves. This threat, never to be taken lightly, may be used almost habitually by a self-punisher who finds that it gets results. Our deep fear is that after years of hearing someone cry wolf, we'll come home someday and find an ambulance crew tending to them.

Eve is a young, attractive artist who is living with Elliot, a prominent painter in his 40s. Their relationship was strong when it began, but once she moved in with him, what had seemed like romantic devotion turned out to be a smothering dependency. She'd seen his mood swings when they were dating, but she'd always attributed them to his "sensitive artistic temperament," and she was completely unprepared to deal with his recurrent depressions, and what appeared to be a possible addiction to sleeping pills. Their relationship has become increasingly distant. The sex is gone, and so is the closeness. Eve works as Elliot's assistant, and he supports her financially, yet he opposes her every attempt to build her own career. He even insists that when she shows her work, it must appear with his.

I finally realized that I have to leave in order to build any kind of life for myself, but every time I make a move to do it, he threatens to take an overdose of sleeping pills. I almost laughed the first time. I said I wanted to take a drawing workshop and he said, "I might as well die then." He's such a drama queen I thought he was teasing. But he keeps saying things like "I can't make it without you" and "If you leave me, I can't promise that I'll go on alone." It's not funny anymore—it's terrifying. I feel such love and empathy for his suffering, but at the same time this rage. Why in the world would he put me in a position like this? All I wanted to do was take a class.

Typical of this form of blackmail, Elliot's threats play on Eve's strong feelings of responsibility. "He's been so good to me. I can't find the strength to leave him. If he did anything to himself, I would never forgive myself," Eve said. And then she added with total conviction, "I would die of guilt."

Most self-punishers don't go as far as Elliot did, though self-punishment can escalate to this level. As I emphasized to Eve, staying with someone is no guarantee that you can save them. Ultimately, the decision to be self-destructive is theirs, not yours. Certainly you can direct a suicidal person to support and resources. But if you remain because you feel that it's your duty to protect them from themselves, you're almost ensuring that whenever they want to strengthen control over you, they will pull out this terrifying form of emotional blackmail.

THE SUFFERER

The image of the sufferer is etched into our culture in a familiar picture: A sour-faced woman sits in a dim apartment, waiting for one of her children to call. "How am I?" she says when the phone finally rings. "You're asking me how am I? You don't call me, you don't visit. You've forgotten your own mother. I may as well stick my head in the oven for all you care about me."

Sufferers take the position that if they feel miserable, sick, unhappy or are just plain unlucky, there's only one solution: our giving them what they want—even if they haven't told us what it is. They don't threaten us or themselves with harm. Instead, they let us know in no uncertain terms, If you don't do what I want, I will suffer, *and it will be your fault.* This last part of the accusation, "it will be your fault," is often unspoken but, as we'll see, can work magic on the conscience of a sufferer's target.

An Award-Winning Performance

Sufferers are preoccupied with how awful they feel, and often they'll interpret your inability to read their mind as proof that you don't care enough about them. If you really loved them, you'd be able to figure out what's bothering them without a single verbal clue. The parlor game they've mastered is "Guess What You Did to Me."

Depressed, mute and often teary-eyed, many sufferers pull away when they don't get what they want, but they don't let us know why. They'll tell us what they want on their own timetable, after letting us twist for hours, even weeks, in anxiety or concern.

My client Patty, a 43-year-old government worker, told me that whenever she disagreed with her husband, Joe, he would dramatically take to his bed.

He'll hardly ever come out and tell me what he wants and on the rare days when he does, if I disagree at all, he gets sad or downcast and goes out for a walk. He has the saddest eyes in the world. We used to have these nondisagreement disagreements when his mother wanted to come over—usually at the most inconvenient time. I gave up fighting that because I felt so guilty when I saw Joe looking so sad.

This is typical. He'll let out an exaggerated sigh, and when I ask what's wrong he gives me a pained look and says, "Nothing." Then I have to figure out what crime I've committed this time. I sit on the bed and tell him I'm sorry if I've done something to upset him, but couldn't he at least tell me what it is? And after an hour or so I'll get the answer to what I did. One time, I had told him I didn't think we could afford the new computer he wanted! How could I be so insensitive and stingy? So of course I told him to go ahead and get it—and surprise, surprise, he cheered right up.

Joe wasn't comfortable sitting down and talking about the computer with Patty. So he made his point another way—he used all the theatrics he could summon to give her the message that she had upset him, made him feel sick, given him a headache. He was flattened by depression because of how "bad" she was. Sufferers look in the mirror and see a victim. They rarely take responsibility for clearing the air or asking for what they want.

Sufferers may look weak on the surface, but they are actually a quiet form of tyrant. They may not yell or make scenes, yet their behavior hurts, mystifies and enrages us.

The Victim of Circumstance

Not all suffering is done in silence. Some sufferers happily share the details of their plights with us and, like their silent cousins, expect us to make everything better. If they're not thriving, we're not giving them something vital to their happiness.

My client Zoe is 57, a striking and confident account executive for a large ad agency. She came to see me because of problems she was having with a colleague at work.

Tess is the youngest person on the staff, and she forgets that it's taken most of us years of doing shitty little jobs and paying our dues to get to where we are. She thinks she should be able to walk right in, with almost 15 years' less experience than the rest of us, and take on the big jobs. I've tried explaining this to her, but the girl's nothing but impatient. Then she started having trouble with our boss and got paranoid about losing her job. Every day she'd come into my office and practically hyperventilate going down the list of what had gone wrong: Dale, one of the partners, hates her copy. She hasn't been able to get through to a valuable client and she thinks he's avoiding her. Her computer doesn't work. And, oh yeah, the dog ate her homework. Sometimes she could see how funny it sounded, but the insecurity was always right there.

She says she's so depressed she can hardly get out of bed in the morning, she's started chain-smoking, she seems to be los-

ing weight. . . . I tried to reassure her, and I thought I was succeeding, but then there was a twist that's making me really uncomfortable. She started pressuring me to put her on my team for a big new project. "I'm going to get fired if you don't," she told me. "Dale hates me, but he trusts you, and if you'd just help me get on his good side, I know that would change everything." Every day it was, "I'm going to get fired if you don't do this teeny favor for me," and "I'm so worried and tormented. You have to help me out."

The fact is, I don't think she's got enough skills yet to pull her weight with the rest of us, but I put her on the project anyway because it seemed almost selfish not to. She really got to me—I bought her idea that it was just me standing between her and a major depression. Not that her job problems had anything to do with her attitude at all. Now I'm worried that I'm going to have to push everyone harder because essentially we're a person short—Tess can't cut it. I felt like such a mentor when I told her she was on. Not anymore. I feel used. You're not going to believe this, but she wants me to give her more responsibility, even though she's struggling with what she's got. I want to help her—I can see my younger self in her in some ways—but this is getting out of hand, and it's going to hurt my own reputation if I don't put a stop to it.

Sufferers like Tess tell us how the cards are stacked against them, the fates have conspired to keep them down. Their theme song might be the old blues tune that goes "If it wasn't for bad luck, I wouldn't have no luck at all." All they need is one little break to turn things around. They often have a certain underdog charm that can be appealing. Of course, these sufferers let us know that if they don't get a break—that is, if we don't give it to them—they'll fail. And that failure, which they can evoke in excruciating detail, will be on our head. They effectively activate our rescuer-caretaker instinct. The problem is, if we give them the "just one break" that they ask for, they'll almost certainly be back for more. Becoming a caretaker to a sufferer is a full-time job, not a stop-gap measure.

THE TANTALIZER

Tantalizers are the most subtle blackmailers. They encourage us and promise love or money or career advancement—the proverbial carrot at the end of the stick—and then make it clear that unless we behave as they want us to, we don't get the prize. The rewards sound juicy, but they turn to dust anytime we get near them. Our desire for what's being promised can be so strong that we endure numerous encounters with forgotten and never-materializing rewards before we realize that we're being emotionally blackmailed.

Over lunch one afternoon, my friend Julie told me about her run-in with a tantalizer, the boyfriend she'd been so excited about the last time we got together. Alex, a wealthy twice-divorced businessman, had been dating Julie, who's an aspiring screenwriter, for seven months. When they met, Julie was running a small freelance writing business from her home and working on screenplays at night. "Your scripts are excellent," Alex had told Julie early on, and, says Julie, he was ever-encouraging.

He said he had a couple of producer friends who were looking for—how did he put it?—intelligent work, just like mine. There was supposed to be a weekend party and he said he'd introduce us. I'd been working so hard, this chance was thrilling. Then came the bait and switch. "Don't invite any of those bohemian friends of yours," he told me. "I think they're holding you back."

When she balked, there were no meetings with Alex's influential friends, just more delicious promises. He gave her expensive gifts—a new computer to replace her ancient word processor, day-care help for her seven-year-old son, Trevor. But each came accompanied by a new hoop for her to jump through. He could open more doors for her if she'd help him arrange more social events at his house. Surely she could give up writing at night so she could entertain with him—it was for her own good.

Attached to Alex, and ambitious enough to thirst for what he had to offer, Julie tried to go along. Then came what was to be the final hoop.

He said he'd been thinking about how much better things would be if Trevor went to live with his father for a while. I'd have more time to work, and I could focus on my career. "It would just be temporary," he said. And then he said something about how I couldn't be on the mommy track when I was so close to making it big.

That woke Julie up, and soon after, she broke off her relationship with Alex. It was impossible for her not to see the relationship for what it was—a never-ending series of tests and demands. A typical tantalizer, Alex was full of gifts and promises, all accompanied by conditions for Julie's behavior: "I will help you if. . . ." "I will ease the way for your career if. . . ." And finally, Julie realized that the testing would never end. Every time she got close to the carrot, Alex pulled it away. Tantalizers offer nothing with a free heart. Every seductively wrapped package has a web of strings attached.

The Price of Admission

Sometimes what tantalizers offer is less tangible than the material rewards Alex dangled before Julie. Many tantalizers traffic in emotional payoffs, castles in the air full of love, acceptance, family closeness and healed wounds. Admission to this rich, unblemished fantasy requires only one thing: giving in to what the tantalizer wants.

My client Jan is an attractive businesswoman in her 50s who's been divorced for eight years and has two grown sons. She's built a successful jewelry business and is now enjoying the fruits of her hard work and creativity. But her relationship with her sister is a great source of pain for her.

My sister, Carol, and I have had a rocky relationship from the get-go. Our parents set us up to be really competitive with one another and were always playing favorites. I was my mother's favorite, and my sister was my father's. But it was

Dad who had the money. He was tightfisted with me, but he always indulged Carol. She knew exactly how to play him. My father was a control freak and couldn't bear to have anyone stand up to him. He set unreasonable rules for us about curfews and dating, and I was constantly locking horns with him. But not Carol. She played the obedient daughter to the hilt, and raked in the rewards. She had a Jaguar for her sixteenth birthday, trips to Europe, the best schools, everything. But she never learned to be self-reliant, whereas I learned early on that if I was going to get anything, I would have to get it for myself.

When my father died, he continued to play favorites even from the grave. He left most of his money to Carol and practically nothing to me. I was hurt and upset when Carol wouldn't share even a small portion of her inheritance, and what little relationship we had deteriorated almost totally. For the next several years we rarely spoke or saw each other, and finally we stopped speaking altogether. The bottom line is that Carol and I just don't like each other very much.

One day last month, out of the blue, I get a call from her. She's crying and she wants to borrow $1,000 from me so they can put food on the table. Her husband is a total screw-up with everything he touches, and he'd lost all their money with crazy investments. Carol pawned her jewelry, and they borrowed money from my mother to keep their house out of foreclosure. It was a mess. Meanwhile, they didn't tone down their lifestyle one bit. They have a valuable art collection, and even a Ferrari.

When Carol saw that I was really resisting, she pulled out all the stops: "I have no one else to turn to." "I don't know what else to do—I thought you could turn to your family when you're in trouble"—suddenly I'm family again.

At first Carol was a classic sufferer, letting Jan know how awful things were for her and that it was within Jan's power to fix them. When she heard Jan's resistance, she switched gears and held out a special carrot.

Her voice suddenly got sweet and she said, "You know, I'd love to have you start coming to dinner and over for the holi-

days. It'll be like the old days." She tapped right into that fantasy I cherish of smiling faces around a beautiful holiday table. My mom is alone now, and I'm not with anyone. Carol's the one who has the intact family, a husband and teenage children. I always get a little sad around the holidays because there's so much distance between us. In my head I know I have friends who I'm closer to than any of my family, but when the Christmas lights go up on Hollywood Boulevard, I get this yearning for the happy family. In my head, I know we've never had that, and we're never going to have it, but in my heart I'd give anything if we could. I have to admit that I was plenty tempted by Carol's "invitation," and I really struggled to figure out the right thing to do.

Carol made it seem that admission to the family fold could be had for a mere $1,000—a minuscule amount for something Jan held so dear. But of course, the price Jan would pay if she gave in to her sister's pressure would be far higher. She'd have to violate her own integrity by enabling Carol to perpetuate her patterns of overspending and financial irresponsibility, and she'd also have to trust someone who had a history of deceiving her.

Yet the temptation Jan felt was real. The fantasy of the good family that Carol conjured for her is hard to resist—a good family is something we all yearn for, something many of us didn't get. The desire for it is strong, and the possibility of getting it pulls us like a magnet. Ultimately, however, I was able to show Jan that if she didn't have the kind of family she wanted by this point in her life, she probably wasn't going to get it. Carol had painted a beautiful picture for her, but it wasn't real. You can't write a check for a thousand or a million or 10 million dollars and buy closeness, no matter what blackmailers might insist.

All the forces that were working on Jan—her guilt; the temptation to look like the successful, adequate one; the tantalizing promise of family that Carol held before her—tapped a deeply vulnerable place within her. But as we'll see, her experience with this profound instance of emotional blackmail was a turning point that helped her find the ability to withstand a coercive form of manipulation.

WHATEVER WORKS

There are no firm boundaries between the styles of blackmail, and as you've seen, many blackmailers combine them, or use more than one. They might take Carol's tack and switch back and forth from suffering to offering up the tantalizing fantasy of a troubled family that could be magically mended—if. . . .

Every style of emotional blackmail wreaks havoc with our well-being. It's easiest to pay attention to the punishers, whose tactics seem the most destructive. But don't for a minute discount the corrosive effects of the quieter types, the ones who are more like termites than tornadoes. Silent or dramatic, both can bring the house down.

Most emotional blackmailers are not monsters. As we'll see, they're rarely driven by malice but rather by some demons of their own, which we will explore in Chapter 5. Because they are usually people who have an important place in our lives, people we may still want to look to as protectors or supporters, I know that recognizing them as blackmailers can be extremely painful. It's not easy to look closely at behavior we may have tried to forgive or ignore and see how it's affected us. But it's a vital step if we hope to put a troubled relationship back on solid ground.

3

A Blinding FOG

Emotional blackmail flourishes in a fog that spreads just below the surface of our understanding, like a bank of clouds under an airplane. As we descend into the blackmail zone, a thick mist of emotions swirls around us, and we lose our ability to think clearly about what the blackmailer is doing, or what we are doing in response. Our judgment becomes hazy.

As I've mentioned, I use FOG as an acronym for fear, obligation and guilt, the three feeling states that all blackmailers, no matter what their style, work to intensify in us. I think it's an apt metaphor for the atmosphere that surrounds all emotional blackmail. FOG is penetrating, disorienting, and it obscures everything but the pounding discomfort it produces. In the midst of FOG we're desperate to know: How did I get into this? How do I get out? How do I make these difficult feelings stop?

We're no strangers to this trio of emotions. We all live with a host of fears, large and small. We all have obligations, and if we have a conscience, we realize that we're interconnected with the people in our lives and have responsibilities to our families and communities. We all live with a certain amount of guilt. We wish we could turn back the clock and undo an action that hurt someone. We regret things left undone. These emotions are central to being in the world with other people, and for the most part, we're able to coexist with them and not let them overpower us.

But blackmailers turn up the volume, blasting us until we're so uncomfortable we will do almost anything—even something that is not in our best interests—to return these feelings to a more tolerable level. Their FOG-inducing tactics prompt responses that are almost as automatic as covering our ears when a siren shrieks past. There's little thinking on our part, only reacting, and that's the key to effective emotional blackmail. When blackmailers pressure us, there is practically no time between feeling emotional discomfort and acting to get relief.

Even though it may sound as if this is a well-thought-out process, most blackmailers create FOG without any conscious design.

FOG sets in motion an elaborate, unseen chain reaction, and before we can interrupt it, we need to understand how it works. The best way to begin is to look closely at FOG's component parts. Though I'll be describing them individually, don't expect these feelings to be neatly separated—they lap over, swirl together and work in combination. Remember, too, that there are as many different sources of fear, obligation and guilt as there are people, and obviously, I can't illustrate every one. The words and actions that trigger these feelings in you may be different from the ones I describe below, but the effects are the same. They all help weave the matrix of discomfort that pushes us to give in to blackmail.

THE REAL F-WORD: FEAR

Blackmailers build their conscious and unconscious strategies on the information we give them about what we fear. They notice what we run away from, see what makes us nervous, observe when our bodies go rigid in response to something we're experiencing. It's not that they're taking notes and actively filing them away for later use against us—we all absorb this knowledge about the people we're close to. In emotional blackmail, fear works a transformation on the blackmailer, too—a process I'll discuss in depth in Chapter 5. In the simplest terms, the blackmailers' fear of not getting what they want becomes so intense that they become tightly focused, able to see the outcome they want in exquisite detail but unable to take their eyes off their goal long enough to see how their actions are affecting us.

At that point, the information they've gathered about us in the course of a relationship becomes ammunition for driving home a deal that's fed on both sides by fear. The terms they offer are tailor-made for us: do things my way and I won't [fill in the blank]:

- Leave you
- Disapprove of you
- Stop loving you
- Yell at you
- Make you miserable
- Confront you
- Fire you

Whatever the specifics, they'll be shaped to fit the contours of the fears we've made apparent. In fact, one of the most painful parts of emotional blackmail is that it violates the trust that has allowed us to reveal ourselves and develop a more than superficial relationship with the blackmailer. In the situations that follow, watch the way the blackmailers zero in on the fears that will get the strongest response.

The Most Basic Fear

Our first encounter with fear comes in infancy, when we literally cannot survive without the goodwill of our caretakers. This helplessness creates abandonment terrors that some people never outgrow. We humans are tribal animals, and the idea of being cut off from the support and affection of those we love and depend on can be almost unbearable. That makes fear of abandonment one of our most potent, pervasive and easily triggered fears.

Lynn, an IRS investigator in her late 40s, married Jeff, a 45-year-old carpenter, five years ago. She came to me because she had accumulated a long list of resentments and bad feelings about him, and she wanted to see if she could find a way to improve their relationship. Jeff quit his job after the two got married. They agreed that because of Lynn's salary, they could afford to let Jeff devote all of his time to maintaining their home, a small ranch near Los Angeles. But the setup is a constant source of friction between them.

Jeff and I don't have the most equal relationship. I make the money and he spends it. No, that's not fair. I work outside the home and he takes care of the ranch—the house, the animals, the property, me. I really like that sometimes, but I'd feel better if he could just make an effort and find some work. The fact is, most of the money we have between us is what I make, but he's always finding ways to run through it, and I'm a pushover when he wants something.

We've been arguing lately about cash flow and our priorities, and in the past few months he's started to sulk when we can't agree—or when I don't agree with him. I'll hear the screen door slam, and he'll yell back something like "I'm going out," then take off for the barn. He knows I can't stand it when he withdraws from me. I'm always following him around the house—if he goes into the other room I practically feel abandoned. When my first marriage broke up, the thing I hated most was the loneliness of coming home to an empty house, and I don't ever want to feel like that again. I've told Jeff about it—and he used to be patient with me, knowing I like to be physically close to him. So it makes me crazy when he stalks off.

The first thing that happens is that I start thinking he's mad at me, so he's going to leave. Intellectually, I know it's crazy. We're having some bad times, but we really love each other, and he's not going anywhere. But this is scaring me. I can't not say what's on my mind, but I'm going nuts with these fights.

Lynn thought of being alone as falling into what she called a black hole, a well of depression that swallows her up when she's by herself. The black hole is the most frightening thing she can imagine, and each time Jeff withdrew, it loomed large in front of her.

We had a major crisis when his old truck died and he started hinting around about how great it would be to have a new one. He could do so much more with a new truck, maybe pick up jobs on the other ranches around the valley. When I told him that I didn't think we could afford it, he was fuming. I hated fighting, but I felt that the money just wasn't there, and I kept telling him so. After a few days of this, he said money was the only thing I cared about, I didn't appreciate all the work he did to make my life pleasant and happy, and maybe I'd appreciate him more if he left me on my own for awhile. Then he took off and didn't come home for four days. I was crazy with worry. I tracked him down at his brother's, and I begged him to come back. He said he wouldn't unless I respected him for what he was and started showing it.

Jeff reacted like a wounded animal, full of defensiveness about his status in the relationship and humiliated by the constant reminders of how financially dependent he was. Despite how much we've evolved socially over the past few decades, relationships like Jeff and Lynn's are still outside the traditional norm, and Jeff, like many men whose partners earn more than they do, felt that he was in a tenuous position—one that he had to justify and protect. The couple had agreed to their financial arrangement, but in Jeff's view, Lynn changed signals whenever he wanted something. Suddenly it was not OK that he was not bringing in any money, and he was off balance, swinging at Lynn to regain his equilibrium.

For her part, Lynn had gone from being confused and scared to a state of panic. Intimate relationships bring up our most potent fears because it is there that we feel the most vulnerable. We may function at a high level in the rest of our lives, only to turn to jelly at any rejection or perceived rejection by a partner.

After all my pleading, Jeff finally came home, but he wasn't talking much, and the tension was so thick I had to do some-

thing. I couldn't stand it. My parents had been like that—distant and angry and silent, all that fake politeness—and I always hated it. I swore I would never live like that with anyone. So I had to cut through the bad feelings. I thought about it, and I asked myself, What's more important—Jeff or the money?

Soon Jeff was driving a brand-new truck. Whether he'd expected the truck or not, now that he had it he felt as though he'd gained a little equality in the relationship, and he knew what it had taken to get Lynn to agree. Though he didn't consciously formulate a strategy of playing on her fear of anger, silence and abandonment, when he felt he wasn't getting what he needed and deserved, he played his trump cards. A pattern developed: Every time Jeff withdrew, Lynn gave in. Jeff had learned that if Lynn became afraid, all he had to do was upset her with his moods and she would give him what she wanted—not that he was a bad guy, not that he was trying to hurt her, but that what he was doing was working.

Because Jeff's blackmail seems to be all about money, Lynn sometimes sounds like an accountant trying fit her feelings onto a balance sheet as she scrambles to avoid facing the terror inside the black hole. And she makes herself crazy obsessing and ruminating:

I'm really nuts about him, but I wonder if I'd be better off without him. Is it just costing me too much money to stay with him? He's completely dependent on me.

She talks more reluctantly about her emotional dependence on him:

How could I think about breaking up and starting all over again with someone else? I'm so afraid of going back into that depressed place I was in before we got married.

I pointed out to Lynn that she was throwing the baby out with the bath water. Yes, there was a financial friction between them, but her fear of abandonment was so blinding that she lost her ability to look at the relationship objectively whenever Jeff blackmailed her. Instead of trying to work out a healthy compromise, Lynn flipped into automatic pilot and resentfully capitulated.

Fear moves us into black-and-white—even catastrophic—thinking. Lynn was sure that if she confronted Jeff, he'd abandon her, and that left her with only two options: give him what he wanted, or choose to break up with him, which would get rid of the blackmail but leave her alone and back in the "black hole." I told Lynn that she had another choice: Together we could deal with the aspect of the relationship that was causing them both so much trouble in the present and work to alleviate her abandonment fears.

Fear of Anger

Anger seems to magnetize fear, pulling it quickly to the surface and activating the fight/flight reaction in our bodies. It's an emotion that few of us can express or experience in a comfortable way because we associate it with conflict, loss and even violence. This discomfort is reasonable and protective, leading us to duck or run when explosive anger threatens to take a physical form and harm us. But in all relationships except significantly abusive ones, anger is just another emotion—not good and not bad. We've built up so much anxiety and apprehension about our own and other people's anger, however, that it can dramatically affect our ability to stand up to blackmail.

For many of us, this emotion seems so dangerous that we're afraid of it in any form, and we fear not only other people's anger but our own. Over the years, I've heard thousands of people express their fears that if they get angry they'll hurt someone, lose control or fall into a million pieces. Just the hint of anger in another person's voice frequently sparks fears of rejection, disapproval or abandonment and, in the extreme, visions of violence or harm.

My client Josh, the furniture maker we met in the last chapter, feels pushed to the wall by his father's angry disapproval of the woman he loves—and frozen by his father's anger. "All I have to do is try to talk to him about it and his whole demeanor changes," Josh said. "I see him tense up, and his voice goes up 20 decibels. When I see that look on his face and hear that roar coming out of his mouth, even though I'm four inches taller, I'm still scared of the guy."

Parents have a remarkable ability to reactivate our childhood fears. As Josh recalled:

When I was a kid, my father used to yell so loud when he got angry that I was afraid the house would come down on us. It's ridiculous, but I still have the same feeling when he's upset with me, though he's mellowed over the years. I act like he's still that scary guy he was when I was a kid.

The events and feelings we experienced as children are alive within us and often reappear when there is turmoil and stress. Though the adult part of ourselves may know that these things took place decades ago, to the part inside all of us that is *not* adult, it's as though they happened yesterday. Emotional memory can keep us locked into old ways of fearful acting and reacting, even when there's nothing in our current reality to justify the fear.

Conditioned Reflexes

Sometimes we react to the mere *suggestion* of the behavior we fear. "All dad has to do is get red in the face and draw his eyebrows together and I back off," Josh said. "It's not like he even has to yell anymore."

Many of you took a beginning psychology class in high school or college, and chances are you remember the story of the Russian physiologist Ivan Pavlov and his experiments with dogs, the classic demonstration of what's known as conditioned reflex. Pavlov was studying the digestive process of dogs, which begins with the natural response of salivating when they see food. He noticed, though, that if he rang a bell as he presented food to them, the dogs soon began to associate the sound of the bell with food and would salivate at the mere sound of it—no food required. In much the same way, the targets of emotional blackmail exhibit conditioned reflexes whenever they have been through an event that's created significant fear in them.

It might be that a husband makes good on his threat to

leave his wife and takes off for a short time; an adult child becomes angry with something her parents have done and stops talking to them for several days; a woman's friend becomes upset, crying and yelling at her. Even if a reconciliation has taken place, the traumatic event is not forgotten. It becomes a symbol of the pain, and by evoking it, a blackmailer refreshes our original feeling of fear, applying enough pressure to make us give in.

For Josh, all it took was an angry look from Paul. He quickly settled on his best course of action: He'd lie. He'd keep seeing Beth but pretend to his father that he'd broken up with her. It was an expedient solution, but Josh's quest to avoid anger cost him dearly. As we'll see people doing throughout this book, Josh is playing a dangerous game called Peace at Any Price. The price for Josh? His own self-respect and the physical and emotional costs of letting anger build, both within him and in his relationship with his father.

Fear flourishes in the dark, unexamined but vividly imagined. Our bodies and the primitive parts of our brains read it as a reason to run away, and often that's what we do, avoiding what we fear because deep inside we believe that's the only way to survive. In fact, as you will see, our emotional well-being depends on doing just the opposite—facing and confronting what we fear the most.

OBLIGATION

We all come into our adult lives with well-established rules and values regarding how much of ourselves we owe to other people and how much of our behavior ought to be determined by such ideals as duty, obedience, loyalty, altruism and self-sacrifice. We all have deeply ingrained ideas about these values, and often we think they're our own ideas, but actually they were shaped by the influence of our parents, our religious backgrounds, the prevailing beliefs of society, the media and the people we're close to.

Often our ideas about duty and obligation are reasonable, and they form an ethical and moral foundation for our lives that I wouldn't suggest being without. But all too often, our attempt to weigh our responsibility to ourselves against the indebtedness we feel to others falls out of balance. We go overboard for the sake of duty.

Blackmailers never hesitate to put our sense of obligation to the test, emphasizing how much they've given up, how much they've done for their targets, how much we owe them. They may even use reinforcements from religion and social traditions to emphasize how much their targets should feel indebted to them.

- "A good daughter should spend time with her mother."
- "I work my ass off for this family, and the least you can do is be there when I get home."
- "Honor (and obey!) your father."
- "The boss is always right."
- "I stood by you when you were going out with that jerk and needed support. All I'm asking now is that you loan me $2,000. I'm your best friend!"

They push far beyond the limits of a give-and-take relationship, letting us know that, like it or not, it's nothing less than our responsibility to do what they ask. These expectations are especially confusing if the blackmailer has been generous to us. But love and willingness can quickly fall out of the equation when replaced by obligation and an enforced sense of duty.

A client I saw several years ago remains in my mind as the epitome of the blackmail target manipulated through obligation and duty. Maria, a 37-year-old hospital administrator who was married to a well-known surgeon, described herself as one of those people who's always there for you. She'd come over at four in the morning if you were depressed and needed company, and she couldn't do enough for the people around her, because she loved the satisfaction of giving.

Her husband, Jay, took full advantage of those traits during their stormy marriage.

I'm from a generation where getting married, having kids and being the good, devoted wife is the most important job a woman can have and Jay probably married me because of it. I love my work at the hospital, but my home—that's the center of my world. I took a seminar at my church that taught me something I always carry with me: It takes one person to make a relationship work. If you give your all and ask for God's help, you can ride with all the waves that come up. As a woman, I'm serious about what I owe my family, and Jay knows that very, very well.

Jay tapped Maria's sense of obligation for years, emphasizing—and probably believing—that no matter what else he did, he was a good provider, holding up his end of the marriage bargain.

People always thought that we were the perfect couple, but what people didn't realize was that Jay was a compulsive womanizer. Before we got married he used to tell me about his sexual exploits and brag about how many women were chasing him or had a crush on him. I didn't really want to hear about them, but it made me feel good to know that with all the women he could have had, he chose me. Now I know just how naive that belief was.

I'm not sure just how many affairs he had after we got married, but I know there were several. Out-of-town conventions, late nights at the office, frequently getting his stories mixed up and his growing indifference to me were all signs. Then came the phone calls from "friends" who saw him with another woman. My gut told me that people weren't making this up, but it took me a long time to confront him. So many conflicting things were racing through my mind. I felt I owed him—he'd worked hard for us.

Jay took the offensive in pressuring Maria to stay with him no matter what—because it was her duty.

He denied everything, of course. "How dare you believe some malicious gossip," he told me. "All I've done is work and sacrifice to see that this family only had the best. There were plenty of times that I didn't want to stay late at the hospital, but I did it for you—and now you're holding it against me. How can you think of leaving and breaking up the family? Just look around at what you have compared to other women. I can't believe you don't appreciate what it took to get this." By the time he was through with me, I had to agree—I did owe him loyalty and trust. And my kids. I love my kids so much. How could I do that to them—they love their dad. How could I split the family apart?

Then he put his hands on my shoulders and whispered in my ear: "Put on that black dress I like so much and I'll take you to dinner. And I don't want to hear the word divorce *again. This is all just gossip that shouldn't concern you." I felt so confused I pasted on a smile, put on that dress and went out with him like nothing had happened.*

Jay knew exactly where Maria's most sensitive spots were, and to reach them he described the consequences of any potential breakup in terms that would speak directly to her sense of duty to her family. Not only would she be abandoning her hard-working husband, he told her, but she'd also be dooming her children to a life of neglect and unhappiness.

Reluctance to break up a family keeps many people in relationships that have gone sour. No one wants to traumatize or uproot their children or have to deal with their pain and confusion. Some blackmail targets feel so passionately about their obligation to their children that they will make what they mistakenly see as a noble sacrifice and give up their right to have a good life. Though Maria was miserable, the thought of shattering her family horrified her, and kept her paralyzed.

Maria's sense of obligation was so powerful that it almost defined her. She was proud of it, and she instinctively defended herself against suggestions that she hadn't lived up to her own standards. Exaggeration followed exaggeration as Jay twisted what duty and responsibility really mean, blowing them up to such a size that they completely overshadowed his own infidelity. As Jay defined it, Maria's obligation to him was all-encompassing. His obligation to her stopped where he wanted it to—in this case, at the idea of remaining faithful to her. In his martyred how-can-you-do-this-to-me stance, he gave little thought to how he could have done this to *her,* and to his children, whose lives had already been affected by the stresses his affairs had brought into the house. How nice it would be if blackmailers were as sensitive to our feelings as they want us to be to theirs.

Jay refused to do anything about his role in the disintegration of their relationship, claiming he was too busy—and there was no need. *He* hadn't done anything wrong, and if Maria was unhappy, *she* should go get "fixed" so they could go back to the way things had been.

I reminded Maria that no matter what position Jay—or for that matter, anyone else—took, she had an obligation to care for herself as well as to look after others. Maria's self-negating willingness to go along with Jay was not coming from a sense of self-respect or an exploration of her choices—it was an automatic reaction to emotional blackmail.

As so often happens with people who are easily manipulated through obligation, Maria remembered to do what was best for everybody except herself. Most of us have a terrible time defining our boundaries, where our obligations to others begin and end. And when our sense of obligation is stronger than our sense of self-respect and self-caring, blackmailers quickly learn how to take advantage.

The Everlasting IOU

Some blackmailers look selectively at the past to find a reason why we owe them whatever they want. Memory, as employed by a blackmailer, becomes the Obligation Channel, with nonstop replays of the blackmailer's good and generous behavior toward us.

When we receive a kindness from a blackmailer, it's not soon forgotten. More like an open-ended loan than a gift, it's always got payments attached—with interest—and we can never seem to get out of the red. The emphasis is on the blackmailer's sacrifices, made not with an open heart but with the intention of scoring points, accumulating markers that can be called in.

When the Target of Blackmail Becomes the Blackmailer

Early in my work with Lynn, I found that she had struck back at Jeff by returning his blackmail with some blackmail of her own.

I asked Jeff to come in for a joint session, and I'll let him describe what happened:

I've gotten to the point where I have to get away from her for a few days. I'd never heard her spell out what she thinks this relationship is about until she called me at my brother's after our fight about the truck. She cried and cried, and finally she started yelling something like "If you really loved me, you never would have done this to me. How can you be so selfish? The only person you think about is you, and it's all take, take, take. You know who makes the money around here. You know who writes the checks. After all I've done for you, how dare you leave me like this. If you ever so much as stop speaking to me again, the money will stop so fast it'll make your head spin." That's when I knew we were in trouble. We were both so scared afterward about what was happening to us that we decided to get into therapy.

Like many blackmailers, Lynn zeroed in on Jeff's sense of how much he owed her, at the same time making a string of negative moral judgments about his character and motives. She did everything she could to coerce Jeff to stay, going beyond emphasizing his obligation to her and trying to make him as afraid as he'd made her. She had given up her power as she frantically hunted for him and begged him to come back, and to regain it she shifted into the blackmailer's role, where she could call the shots.

It's not uncommon for two people in any relationship to change roles, alternately playing both target and blackmailer. One person may blackmail more than the other, but rarely is blackmail completely one-sided. We may be the target of blackmail in one relationship and turn right around to become the blackmailer in another. For example, if your boss is using emotional blackmail on you at work, the frustration and resentments you feel and either can't or won't express directly to him may cause you to use the same tactics on your partner or children to regain some sense of control. Or, as in Lynn and Jeff's case, the shift may occur within one relationship, with the target changing hats and blackmailing the blackmailer.

Obligation is a particularly tough feeling to keep in proportion in our lives. Too little and we shirk our responsibilities. Too much—as when Lynn started "billing" for every contribution to the relationship—and we're pressed flat beneath the weight of inescapable debts and the inevitable resentment they produce. Blackmail is quick to follow.

GUILT

Guilt is an essential part of being a feeling, responsible person. It's a tool of the conscience that, in its undistorted form, registers discomfort and self-reproach if we've done something to violate our personal or social code of ethics. Guilt helps to keep our moral compass working, and because it feels so painful, it dominates our attention until we do something to relieve it. To avoid guilt, we try to avoid doing harm to someone else.

We trust this active gauge of our behavior, and we believe that whenever we feel it, it's because we've stepped out of bounds and willfully violated the rules we've set for ourselves about what's acceptable in our dealings with other people. Sometimes that's true, and our feelings of guilt are a natural, appropriate response to having done something hurtful, illegal, cruel, abusive or dishonest.

Guilty feelings are woven through our lives if we have a conscience. Unfortunately, our sense of guilt can easily give us false readings about the impact of our actions. Like an oversensitive car alarm that's supposed to alert us to a crime but winds up going off whenever a truck rolls by, our guilt sensors can go haywire. When this happens, we experience not only the appropriate guilt I described above but also what I call undeserved guilt.

In undeserved guilt, the remorse we feel has little to do with identifying and correcting harmful behavior. This kind of guilt, which makes up a large part of the blackmailer's FOG, is layered with blame, accusations and paralyzing self-flagellation. In simple terms, the process that produces undeserved guilt looks like this:

1. I act.
2. The other person gets upset.
3. I take full responsibility for the other person's upset, whether I had anything to do with it or not.
4. I feel guilty.
5. I will do anything to make reparations so that I can feel better.

More specifically:

1. I tell a friend I can't go to a movie with her this evening.
2. She gets upset.
3. I feel horrible and believe it's my fault she's upset. I feel like a bad person.
4. I cancel my other plans so we can go to the movie together. She feels better, and I feel better because she feels better.

Undeserved guilt may have nothing at all to do with our harming someone else, but it has everything to do with *believing* that we did. Emotional blackmailers encourage us to take global responsibility for their complaints and unhappiness, doing all they can to reprogram the basic and necessary mechanisms of appropriate guilt into an undeserved-guilt production line where the lights continually flash guilty, guilty, guilty.

The effect is strong. All of us want to believe that we're good people, and the guilt that blackmailers evoke attacks our sense of ourselves as loving, worthwhile people. We feel responsible for the blackmailers' pain, and we believe them when they tell us that we're compounding their misery by not accommodating their wishes.

The Name of the Game Is Blame

One of the fastest ways for blackmailers to create undeserved guilt is to use blame, actively attributing whatever upset or problems they're having to their targets. Since our guilt system is put on alert by situations that cause us to ask, Have I harmed someone?, most of us feel pangs of guilt anytime someone directly accuses us of hurting them—whether we've done anything to deserve the feeling or not. Sometimes we're able to short-circuit the guilt reaction by seeing that there's no connection between the accusation and reality, but many times we apologize first and double-check the blackmailer's logic later—if we check it out at all.

We often talk about guilt-peddling, but I think it's more accurate to talk about blame-peddling. Like turn-of-the-century pushcart salesmen, emotional blackmailers who peddle blame bombard us with a sales pitch aimed at getting our attention. Though the details vary, one phrase, often unspoken but always just under the surface, serves as the blame peddler's slogan: *It's all your fault.* It's the hook that makes us buy what they're selling.

- I'm in a lousy mood (and it's all your fault).
- I have a bad cold (and it's all your fault).
- I know I drink too much (and it's all your fault).
- I had a bad day at work (and it's all your fault).

When you see a list like this, the pitch sounds absurd. Chances are good that these complaints had *nothing* to do with the much-maligned you. But often we don't recognize the confusing messages for what they are, because most of us have a tendency to take the blame for something when a person we care about is upset. The blackmailer is only too happy to explain to us how and why we bear all the responsibility for a situation and why they bear little or none. We buy the blame, and our guilt flows freely. We're ripe for the relief that comes from giving in to the blackmailer.

ENTWINING THREADS

It's impossible to separate the emotions that form the FOG within us as emotional blackmailers set the stage for manipulation. Where you find one element of the FOG, the others are almost certainly close by.

With Maria, for example, obligation and guilt were tightly bound together. Few of us can contemplate not fulfilling what we consider to be our obligations without feeling guilty, and Maria was no exception.

Jay hammered home to me that if we broke up, it would be totally my fault. I'd lie in bed thinking about what it would mean to fail as a wife and mother, and I felt guilty as hell. I'd have to say I was almost paralyzed for a long time. I couldn't bear the thought of letting down the kids—my God, they don't deserve to have their lives torn apart or ruined. Anything good I'd ever done seemed to be completely erased by the thought of breaking up the family. I could hardly speak the word divorce *because it made me feel so selfish.*

Once again, Maria put herself last, and Jay knew that he could count on her to keep doing it. Although Jay's actions gave her good reason for feeling angry and wounded, those feelings were overshadowed by her ever-growing sense of guilt.

Like Maria, many people go through the motions of everyday interactions with a blackmailer who's plying them with guilt, but the resentment and even self-loathing that build in them are corrosive. With little fun or true intimacy, what looks like a marriage or a friendship becomes a hollow framework.

NO STATUTE OF LIMITATIONS

Once blackmailers see that their target's guilt can serve them, time becomes irrelevant. If there's no *recent* incident to peg their guilt and blame-peddling on, one from the past will do just fine. There's no point at which a guilt-producing event is allowed to diminish and reparations are considered fully made. Blackmail targets discover that whatever their real or imagined transgression may be, there is no statute of limitations—no point after which an ancient "crime" ceases to be an issue, or a punishable offense.

Karen, the nurse we met in Chapter 2, was kept in a fog of guilt by her daughter, Melanie, who never let her forget an accident that occurred in her childhood.

It goes back a long way. My husband, Melanie's dad, was killed in a car accident when Melanie was little. Melanie was in the car—she was seriously injured and her face was scarred. I paid for plastic surgery for her and she looks fine, but she's still self-conscious about a few remaining marks on her forehead. And of course I paid for years of therapy because I know how tough all of this was for her.

It's taken me a long time to work through my guilt about that night. I know it was the other driver's fault—but if only we hadn't turned down that street . . . If only we had waited until the next day to leave the way my husband wanted to. . . . If only . . . Melanie has her own version of why I'm the villain of the story. She never fails to remind me that we were going on a camping trip because I insisted I needed some R&R. If I hadn't just been thinking of myself and getting away from the office for a few days, the car wouldn't have been where it was and the accident wouldn't have happened. I know it's irrational, but it feeds right into my own guilt and I end up giving her whatever she wants.

No matter what Karen did to ease the sting of her guilt, however, Melanie never let her forget it for long. Karen found, as all blackmail targets inevitably do, that giving in once or twice doesn't end the blackmail, it only intensifies the demands.

Sometimes I wonder—do I have to keep making up for this for the rest of my life? I've tried to help her, but nothing's enough. I know I'm not to blame for her troubles, but everything seems to come down to that one moment when some drunken bastard rammed into our car.

Karen's guilt is fused with a sense of obligation to her daughter. To Karen, this continuing sense of guilt means that she'll always owe Melanie for what happened to her, even though it was not Karen's fault, and until Karen sees what's going on, she will always capitulate to Melanie's demands in an effort to repay her for her pain.

WHEN APPROPRIATE GUILT GETS OUT OF HAND

Even if the guilt we feel is appropriate, an emotional blackmailer will not let us forget what we've done or allow our guilt to serve its function of helping to correct our behavior and serving as a teacher for the future. Bob, the lawyer we met in Chapter 1, knew that he'd violated the trust of his wife, Stephanie, by having an affair, and he wanted desperately to make amends and find a way to allow the relationship to heal. But Stephanie had been badly hurt, and she insisted on reminding him of his offenses. Bob struggled to deal with what had now become lingering, inappropriate guilt.

I don't know what else I can do to make things up to her. I have to go out to make a living, and I can't spend all my time reassuring her. I don't know how to make her feel safe again, and she won't let me know what it'll take. She won't let go of it, though. I made her suffer, so she's going to make sure I suffer as much or more. Jesus, even criminals get out of jail eventually, but I'm doing life with no chance of parole.

Stephanie had every right to be angry and hurt, but she was keeping herself and Bob suspended in time and using his guilt to control him. As long as guilt dominated their interactions, there was no chance for healing. Until Stephanie and Bob both learned how to handle this volatile emotion, neither of them was able to see a way out of the emotional blackmail that had put their marriage in a deep freeze.

Guilt is the blackmailer's neutron bomb. It can leave relationships standing, but it wears away the trust and intimacy that make us want to be in them.

BEFOGGED AND BEWILDERED

Many years ago, I lived in a beach community where, several times during the year, the fog rolls in late in the day and lingers through the night. One evening, as I was coming home late from work, the fog was particularly heavy, and I drove through my neighborhood straining to see. I was relieved when I reached my street and saw my driveway. But for some reason, I couldn't get my garage door opener to work. When I got out of the car to check it, I realized that I'd pulled into the driveway of the house next door. I simply couldn't see what I was doing until I'd done it.

My experience that night was identical to what happens when we travel through the FOG of emotional blackmail. Even if our bearings are good, the FOG that emotional blackmailers create adds a new dimension that disorients us in the midst of even the most familiar situations and relationships.

We can't have emotional stability if FOG controls our lives. It dismantles our sense of perspective, warps our personal histories and muddies our understanding of what's happening around us. FOG bypasses our thought processes and goes straight for our emotional reflexes. Suddenly we're floored, and we don't know what hit us. Score: blackmailer 100, target 0.

4

Tools of the Trade

How do our blackmailers create FOG in our relationships with them? How do they maneuver us into putting aside our own best interests as we fall into the frustrating pattern of demand-pressure-compliance? We can begin to see how the process works by looking more closely at the strategies blackmailers consistently use—their own specialized tools of the trade.

Singly and in combination, these devices heighten one or more elements of the FOG, magnifying the pressure within us to find relief by saying yes to the blackmailer's requests. They also help blackmailers justify their actions, both to themselves and to us. This element is crucial, because it helps cloak the blackmail in acceptable, even noble, reasons for being. Like parents who punish their children while saying "I'm only doing this for your own good," blackmailers are expert rationalizers, and they use their tools to persuade us that the blackmail somehow serves us.

The tools are a constant that runs through the endlessly varied scenarios of emotional blackmail, and all blackmailers, no matter what their style, use one or more of them.

THE SPIN

Blackmailers see our conflicts with them as reflections of how misguided and off-base we are, while they describe themselves as wise and well-intentioned. In the most simplistic terms, we're the bad guys, and they're the ones in the white hats. In politics, this process of running events through the good-guy/bad-guy filter is called "spin," and emotional blackmailers are the original spin doctors, masters of putting a halo around their own character and motives and splattering ours with serious doubts, or even blacker mud.

The Spin Doctor

I had a call one day from a woman named Margaret, who told me that her marriage was in serious trouble and wanted to see if there was a way to salvage it. We made an appointment, and when she came in, I was struck by her charm and graceful bearing. Margaret was in her early 40s and had been divorced for five years when she met her new husband at a singles group of the church they both attended. After a short, intense courtship, Margaret and Cal got married. They'd been together about a year when she came to see me.

I'm so confused and depressed. I need some answers here—I don't know if I'm right or he is. I really thought I'd hit the jackpot this time. Cal is good-looking, successful and I thought really kind and caring. The fact that we met at church was important to me because it meant we had similar values and beliefs. So imagine my shock when about eight months into the marriage he announces that he wants me to participate in group sex with him—and he's been doing it for years. He said he loved me so much he wanted to share it with me.

I told him there was no way I would do such a thing—the idea revolted me—and he acted genuinely shocked. He said he'd always loved my sensuality, and he wanted to introduce me to something that would really enrich my life. He said that he knew he was taking a risk by bringing it up, but it was proof of his love for me that he wanted to share everything with me. And doing it with him would be proof of my love for him.

When I said no way, he acted really hurt and kind of angry. He said he'd badly misjudged me. He thought I was liberal, open-minded and loving, and he had no idea I was such a prude and a puritan—that wasn't the kind of woman he had fallen in love with. And then he really stuck the knife in. He said that if I didn't do it, there were plenty of his old girlfriends who would.

Like all spin doctors, Cal was interpreting his desires in glowingly positive terms and describing Margaret's resistance as darkly negative. Blackmailers let us know that they ought to win because the outcome they want is more loving, more open, more mature. It's what's best. They are entitled to it. At the same time—and sometimes in the most polite way—they call us selfish, uptight, immature, foolish, ungrateful, weak. Any resistance on our parts is transformed from an indication of our needs to evidence of our flaws.

Cal even implied he had been misled or deceived by Margaret's earlier behavior. But she could change that assessment by going along with him, an act that would prove she was the open, sensuous woman he wanted her to be.

Confusing Labels

I'm focusing on the labels Cal used on Margaret because the spin involves applying adjectives—positive ones to the blackmailer and the compliant target, negative ones to the person who resists. Cal interpreted his differences with Margaret as an indication that something was wrong with her and then proceeded to use labels that reinforced his position. The experience is disorienting. Blackmailers' labels are so different from the ones we'd choose that before long we doubt the labels we give to things ourselves, and we begin to internalize the blackmailers' questions about our perceptions, our character, our worth, our desirability, our values. We're trapped in a dense FOG of the worst kind, as Margaret told me:

I couldn't make myself see how Cal could be so different from the man I thought I'd married. How could I have been so wrong about him? I couldn't believe it. In the most rational way you can imagine, he made it seem as though I'd led him to believe I'd do anything with him, and he kept saying how good it would be for us as a couple. It was easier for me to think I was missing something and that if I could just understand Cal's ideas about group sex it wouldn't seem so outrageous to me. I really struggled with it. I thought, Maybe I am *uptight. Maybe I* am *a little prudish. Maybe I just don't get it. I began to think there was something wrong with me and I'd been making a big deal about nothing.*

Margaret had been certain that for her—and the marriage—group sex couldn't have any positive effects, but as Cal kept talking, she began to doubt herself. When the spin is effective, it confuses us about what's harmful or healthful and makes us question what we see going on between ourself and the blackmailer. We buy into the spin because we want our friends, lovers, bosses and family members to be right and good, not mean, unfeeling or oppressive. We want to trust the other person instead of acknowledging that he or she is manipulating us by labeling us in ways that make us feel ashamed or inadequate.

Margaret tried hard to make the situation fit logically into the picture she had envisioned of her life with Cal. Surely there was something she hadn't yet comprehended, an interpretation that would make Cal's demand seem acceptable. If her concerns were valid, what did that say about their marriage, about him? These were frightening questions, and on some level Margaret didn't want to face them. She didn't want to admit to herself that she'd made a mistake about him. It was less painful for her to buy Cal's version of reality than to confront the uncomfortable truths about him and their relationship.

Cal, as well as inducing Margaret's self-doubt, leaned heavily on her sense of obligation. In his spin, it was her duty as a wife to participate in group sex with him—he didn't want a wife who denied him that. Imagine how surprised and vulnerable she must have felt when he threatened to replace her with someone who would accommodate his "reasonable" request.

Unfortunately, Margaret gave in.

I can't believe I buckled under to his pressure and agreed to try it if it meant that much to him. I'm so ashamed. I hated every minute. I feel dirty and angry and terribly depressed.

The FOG was so thick, and Margaret had been pushed so far off center, that it was hardly a surprise she wound up acting in ways she never would have considered otherwise.

Making Us "Bad"

In addition to discrediting the perceptions of their targets, many blackmailers turn up the pressure by challenging our character, motives and worth. This type of spin tactic is popular in family conflicts, especially those in which parents are clinging to control of their adult children. Love and respect are equated with total obedience, and when that's not forthcoming it's as though a betrayal has taken place. The party line of the blackmailer, repeated with infinite variations, is *You are only doing this to hurt me. You care nothing for my feelings.*

When Josh fell in love with Beth and began thinking of marrying outside his religion, he knew his parents would be upset, but he didn't count on the all-out attack his father launched in an effort to get him back in line.

I couldn't believe what Dad was saying. You'd think I had created some kind of conspiracy to wreck his life. Why was I torturing him? Why was I driving a spike into his heart? Overnight I went from being a good son to the biggest screwup in the family.

Josh had been away from his parents for several years, but like most people who hear a parent say "You've hurt me," or "You've disappointed me," Josh felt the impact of those statements like a punch in the stomach.

Such words, coming from someone we're close to, cling to us, affecting the inner gyroscope that guides our actions and leaving us with a wobbly sense of ourselves. Obviously, we might be labeled heartless, worthless or selfish in any relationship with a blackmailer, but those labels are especially difficult to withstand when they're coming from a parent, whom we spent our formative years regarding as a repository of wisdom and righteousness. Parents who use the spin against us can wipe out our confidence almost faster than anyone else.

PATHOLOGIZING

Some blackmailers tell us that we're resisting them only because we're ill or crazy. In the therapy business, this is called pathologizing, and much as I hate using psychiatric jargon, the word *pathologizing* describes this activity to a T. *Pathology* comes from the Greek word *pathos*, which refers to suffering or deep feeling, but the current connotation of the word is "disease." Pathologizing is a way of making us appear "sick" when we don't go along with a blackmailer. Blackmailers accuse us of being neurotic, warped, hysterical. And most wrenching, they dissolve the trust that's accumulated in a relationship by lining up all the unhappy events we've shared with them and throwing them back in our faces to prove that we caused these things to happen because we were such emotional cripples.

Because the experience of being pathologized by blackmailers can be a devastating blow to our confidence and sense of self, it's an especially toxic—and effective—tool.

When Love Is a Demand

Pathologizing often arises in love relationships when there's an imbalance of desires. One person wants more than the other—more love, more time, more attention, more commitment—and when it's not forthcoming, he or she tries to get it from us by questioning our ability to love. Many of us will go to great lengths to prove that we are loving and lovable, and many of us have the erroneous belief that "If somebody loves me, I'm supposed to love them, too—or there's something wrong with me."

My client Roger, a screenwriter in his mid-30s, faced a blizzard of pathologizing when he decided he needed some independence from Alice, an actress he'd met at a 12-Step meeting eight months earlier.

I have the feeling that Alice is more devoted to me than anyone I've ever known. It was an incredible high to spend time with her right after we met. She'd come over and sit on my bed reading drafts of my work and raving about them. She seemed to appreciate what I was trying to do, and to love it as well as me. I fell for her hard. She's seen every movie ever made, she's funny, she's gorgeous and she thinks we were meant to be together.

But after only a couple of months, she was pushing for us to live together. She kept saying how thrilled she was that we'd found each other and that she knew we'd transform each other's lives. All I needed to do was surrender my resistance—let go and let God guide us into a great relationship. She said she knew I might be reluctant because of the bad breakup I'd had with my ex last year, but I had to face my fears, not run away from them. It sounded good, but it felt way too fast.

Alice and Roger spent a lot of time talking about the personal 12-Step work they were doing, and they supported one another with it. But Alice liked to play therapist, especially when Roger talked about his fear that their relationship was moving too fast. He was trying to control things, she told him, and he needed to stop resisting. Even at this early stage, Alice tended to define Roger's hesitation as leftover neurotic behavior from his drinking days, though he'd been sober for eleven years. And Roger took her feedback to heart. Despite the nagging sense that he was in over his head with Alice, he decided that maybe she was right. He told Alice she could move in.

She was so clear about our future, and I was just trying to take things one step at a time—but when someone loves you like that, they seem to get this huge surge of energy, and you get swept up in it. I'll admit it made me kind of tense, but I'm coping. In the past couple of months, though, she's started talking about having a baby. She's 35, and the woman has got serious baby lust. She says we wouldn't have to get married, but this would be the perfect chance to express all of our love and creativity. She's been reading me baby books and pulling out pictures of me as a kid to see what the baby will look like. It's too much. I don't know if I want to spend the rest of my life with her, or if I ever want to be someone's father. I need space to work and write.

It's not that I don't like her and think she's great, but I've got to sort things out. I'm not sure I feel for her what she seems to feel for me—I'm just not sure. So I said I needed to live by myself for awhile to get some perspective.

Roger's resistance to Alice's plans triggered a furious response.

She said something like "I'm afraid of you when you talk this way. You said you loved me, but from what you just told me, I have to think that you're a fabulous liar. I know you're afraid of getting close to me after the mess you made of your last relationship, but I thought you were ready to start living in the present instead of the past. I know I'm an intense person, but I thought I'd met my match. I guess I can't be mad at you, but I pity you. You're too afraid of life to ever experience love. You're only comfortable with your little screenplay fantasies. Face it, you're a dry drunk just like your womanizer father."

Then, with a nervous laugh, Roger said:

I keep replaying it in my mind and wondering if she's right. I do have a rough time establishing relationships. Maybe I don't know how to be with someone who really loves me.

I told Roger that he'd overlooked something many people forget: There's nothing "wrong" with you just because you don't want someone as much as they want you. Like many pathologizers, Alice misused the word *love.* Her actions were full of dependency, desperation and the need to possess Roger totally, none of which has anything to do with mature love. But to her the pressure was justified in the name of her grand, overpowering passion for him—and if Roger couldn't match her intensity, the only explanation she could live with was that he must have some horrendous psychological problem.

As she reacted to Roger's request for more space, Alice used a tactic that most pathologizers rely on: throwing back in Roger's face uncomfortable things that he'd confided to her about himself and his family. Roger had told her about his dad, who had given up drinking only to replace it with a compulsion for womanizing, and Alice knew that, like many of us, Roger strongly feared "turning into my father." Secrets, fears or confidences we've shared with a pathologizer become easy-to-grab weapons in a conflict. Painful life events—divorces, child custody battles, abortions—that we have described at intimate moments are all used as proof of our instability. For Roger, Alice's "evidence" that his hard-won sobriety was somehow tainted was cause for alarm.

Emotional blackmailers often accuse us of being unable to love or maintain friendships simply because we don't feel as close to them, as friends or lovers, as they feel we should. It's a variety of pathologizing that many of us are vulnerable to especially if we see intimate relationships as the litmus test of mental health. Though it's a stretch for blackmailers to argue that if a relationship isn't working, it's because we are sick or damaged, lines like this go straight to the heart, and often they succeed.

What's Wrong with You?

Not all pathologizers overtly label a person as sick. This tool has more subtle guises. My client Catherine sought me out, her confidence shaken, after having repeated run-ins with her former therapist.

I was ready to go part-time at my accounting job and work on getting my MBA, and I was feeling pretty anxious about that. But mainly, I had recently had a bad experience with a guy, and I wanted to understand what happened. So I decided to go see a therapist that my friend Lanie was enthusiastic about.

There was something forbidding about Rhonda from the first, but I figured it would just take time to get used to this new kind of relationship. It seemed to me that she was always getting in little digs at me, though. One of her favorite things was to clip articles out of the paper about successful women who had it all and give them to me at the beginning of the session for "inspiration." It made me feel like shit. The message was, "Here's the way you should be, and if you do what I tell you to do, you'll get there."

She kept bringing up my joining one of her therapy groups, but I didn't have any interest in that. Maybe she was right that it would do me a world of good, but my god, I had so many hours to put into my master's program and working that I didn't have any time. Rhonda saw it another way. She said that I was being stubborn and controlling and that was why I was having problems in my life.

Pathologizing is especially persuasive when it comes from an authority figure—a doctor, professor, lawyer or therapist. Our relationships with these people are based on trust, and we tend to cloak professionals in a mantle of wisdom that some don't deserve. We assume that they will treat us with openness and integrity. But we've all met authorities who seem to believe that their license to practice puts their opinions and actions beyond reproach. They may never come out and say "You are deficient," but with a gesture, a harsh or critical tone of voice or a set jaw, they convey that you are and that your position is wrong.

It was clear from her tone, her body language and her whole attitude that she was upset with me—and that felt awful. I was afraid she might get mad at me. And that would be the final confirmation of the fact that I wasn't OK. After all, your therapist is the arbiter of what's right and wrong, and if your therapist doesn't like you or disapproves of you, then there's really something wrong with you. Also, I've always been scared of anger, harsh words. When you're dealing with someone in a position of authority, that gets multiplied by ten.

Arrogantly, "authorities" like Rhonda indicate that they are not to be challenged. They tell us that they're looking out for our best interests, and by resisting them we're proving how obstinate, ill-informed or insecure we are. *They* are the experts, even when it comes to our deepest knowledge of ourselves, and we aren't allowed to question their advice or their interpretations of a situation.

Dangerous Secrets

Many families that have "shameful" secrets of child abuse, alcoholism, emotional illness and suicide tacitly agree to keep the facts hidden and never discuss them. But when one person changes the signals by pulling out of the family system that survives on denial and secrecy, it's typical for family members to brand that person crazy, unforgiving or a family wrecker for daring to discuss their hidden and long-denied history. I saw this kind of pathologizing frequently during the years I specialized in working with adults who had been sexually and/or physically abused as children. As my clients began to get healthier, they wanted and needed to talk about their experiences, but some families fought hard to keep them from breaking the silence.

It's axiomatic that the more troubled the family, the more it tries to block its members when they try to become healthy again. And all too often, the blackmail works. Threats of abandonment, exile, punishment, retaliation and total disapproval or contempt can crush the resolve of the person whose brave attempt to recover has been pathologized as selfish, unnecessary and destructive.

Roberta, a 30-year-old telemarketing executive, still lives with injuries to her neck and bones that she suffered as a child at the hands of her abusive father. When we met, she had been hospitalized for depression in a facility where I was on staff. One of the first things she told me is that she couldn't bear to keep holding in the family secret of abuse.

As Roberta began to deal with her childhood, she looked to her mother for confirmation of what she'd seen and experienced, but instead of the understanding she had hoped for, she was met with pathologizing.

I tried approaching my mom about six months ago and telling her that I found out that I still have some old injuries from Dad's beatings. And she totally blew me away. She told me I was making it sound like my dad had killed me or something. And I said, "Do you remember when Dad got me by the hair and swung me around and threw me on the ground?"

She looked at me as if I was an alien from outer space. She said, "Oh my God, where are you getting all this from? What are those doctors telling you over there? Have you been brainwashed?" And I said, "Mom, you were standing there most of the times when I got hit—in the doorway watching." She flipped out. She could not handle it. She said I was making everything up and she thought I was out of my mind. How could I talk about my father like that? She said she couldn't stand to talk to me until I got help and stopped needing to tell such terrible lies. It was devastating.

Roberta's mother found her daughter's memories so threatening that she not only denied them, she also pressured Roberta to deny them herself and threatened to cut off contact until Roberta stopped upsetting the family. Healthy attempts like Roberta's to disclose and discuss what had happened to her are frequently given an almost sinister cast by other family members, labeled "fiction" and "exaggeration" and the product of a sick mind. We may desperately need to express the truth of what's happened to us, but it takes determination, preparation and support to counter the pervasive pathologizing that goes along with long-term abuse or deep problems within a family.

Pathologizing targets an area that's hard for us to defend. It would be easy for most of us to defend ourselves against criticism of our skills and accomplishments because we're surrounded by solid external confirmations of what we can do. But when a blackmailer tells us that we are somehow psychologically deficient, we may accept their description as rational feedback. We know we can't be completely objective about ourselves, and many of us are terrified that we have unknown demons inside. Pathologizers play on that fear.

Like the spin, pathologizing makes us unsure about our memories, our judgment, our intelligence and our character.

But with pathologizing, the stakes are even higher. This tool can make us doubt our very sanity.

ENLISTING ALLIES

When single-handed attempts at blackmail aren't effective, many emotional blackmailers call in the reinforcements. They bring in other people—family members, friends, ministers—to help make their case for them and to prove that they're right. In this way, they double- or triple-team their targets. Blackmailers round up anyone they know the target cares about and respects, and faced with this solid front, the target may feel outnumbered, overpowered.

One evening shortly after I began working with Roberta, I saw this tool in action. Roberta's parents showed up for a family counseling session with Roberta's brother and two sisters, all of whom were eager to show their solidarity with their parents. When I asked them how they felt about Roberta's desire to speak openly about her father's abuse, I noticed how they closed ranks. The siblings looked at each other, and finally her older brother, Al, began to speak.

Mom called and asked us if we'd come in and let you know the truth about what happened in our family because we're a good family, and Roberta's just trying to destroy us. You know how sick she's been—look at her, in and out of the hospital for depression, suicide attempts. . . . It wouldn't surprise me if she's hearing voices or something.

He smiled, looking around the room as his parents and siblings nodded.

She's always had big problems. We all want to help her get better, but we can't let her tell horror stories about us. She's made up this abuse thing, and a lot of people seem to believe it. We just want to clear our name and see that she gets the help she needs.

Roberta had had a hard enough time holding on to what she knew in the face of her mother's denial, and now her task was even more difficult. She was facing a whole roomful of blackmailers who wanted her to keep her mouth shut. And their full-court press let the "traitor" know that she would be welcomed back into the fold only if she'd be quiet, so everyone could resume a way of behaving that, despite being destructive to all of them, was familiar, and therefore comfortable.

Bringing in Fresh Troops

My client Maria, the hospital administrator we met in the last chapter, provides another compelling example of this kind of double-teaming. When she discovered her husband's extramarital affairs and told him that she was thinking of leaving him, he tried everything to persuade her to change her mind—including enlisting his relatives.

He saw that threats and charm and all the things that had worked for him in the past weren't working anymore, so he brought in the heavy artillery—his parents. I was crazy about my in-laws. His father was a doctor, too, and his mother was a gentle soul who had been wonderful to me from the day I met her. So when Jay's dad called and suggested that we have a family conference at their house, I was hesitant about it, but I felt I owed him that consideration.

From the moment I walked into their house, I knew I'd made a mistake. Jay had gotten there early and had obviously been talking about how unreasonable I was being. How could they possibly be objective about their golden boy, and how could they be fair to me?

Maria's concern was well founded. There was no way that Jay's parents could be objective in this situation, and what happened next was no surprise.

For over an hour, they talked about how all marriages go through bumpy times, and you can't walk away at the first sign of trouble. They said Jay had agreed to spend more time at home and to cut down his hours at the hospital, and that should solve our little disagreement. Now if I would just stop using the word divorce, *nobody would have to know anything about it. They asked me if I really wanted the breakup of the family on my conscience, especially knowing how much Jay loved me. They said it was breaking their hearts to see him suffering, and didn't I know what this would do to the children? How could I bear to make so many people unhappy when my husband was doing so much to provide a good future for me?*

When I asked them if Jay had told them about his extramarital affairs, I could tell from their reaction that he hadn't. They looked so uncomfortable I thought maybe they'd understand a little better why I'm so unhappy being with their son. Then his father said the most incredible thing. He said, "That's not a good reason to break up a family! Family comes first. You can't break down at the first sign of trouble and throw it all away. Think about the kids—our grandchildren." That stung!

Now, instead of having one person twisting her arm, Maria had three, and it took all her inner resources to hold her ground against them. The message from all of them was the same—it was as though Jay himself had written the script—but hearing Jay's words from other people she loved and trusted made those words carry even more weight.

Calling on a Higher Authority

When friends and family aren't sufficient reinforcement for emotional blackmailers, they may turn to a higher authority such as the Bible, or other outside sources of knowledge or expertise, to shore up their positions. This form of pressure may sound as simple as "My therapist says you're being bitchy" or "I took a course where they told us . . ." or "Dear Abby says. . . ."

Wisdom resonates differently in each of us, and none of us can claim a monopoly on it, but we can count on blackmailers to insist, by pulling selective quotes, comments, teachings and writings from a host of sources, that there is just one truth: theirs.

NEGATIVE COMPARISONS

"Why can't you be like . . ." Those five words pack an emotional punch that connects powerfully with our sense of self-doubt, our fear that we don't measure up. Blackmailers often hold up another person as a model, a flawless ideal against which we'll fall short. *That* person would have no problem satisfying the blackmailer's demands—so why can't we?

"Look at your sister—she's willing to help with the business."

"Frank doesn't seem to be having any trouble meeting deadlines—maybe you could get some pointers from him."

"You don't see Mona abandoning her husband when the going gets rough."

Negative comparisons make us feel suddenly deficient. We're not as good, not as loyal, not as accomplished as so-and-so, and we feel anxious and guilty about it. So anxious, in fact, that we may be willing to give in to blackmailers to prove that they're wrong about us.

My client Leigh is a stockbroker whose mother, Ellen, has a black belt in negative comparisons, and Leigh has felt the pressure in a variety of forms over the years.

When my dad died, Mom was totally helpless. All her life she'd had men taking care of her, and when Dad died she turned to me to take over her life for her.

I found out fast that I was supposed to (a) spend an enormous amount of time with Mom and (b) find her a lawyer, find her an accountant and do a lot of things that she was perfectly capable of doing for herself. But she's good at playing helpless, and I immediately jumped in. I have an easy time doing all those things, so it wasn't a problem for me. The myth is that you're going to get love and approval for this. But what really happens is that you can't do anything to please a woman like her. So of course the accountant charged her too much, and the lawyer was a no-good so-and-so, and I was a criminal for having to skip dinner with her because I promised my son I'd coach him for the play he's in.

If what I did wasn't exactly perfect, I sure heard about it. And anytime I started to pull away, she started leaning on my cousin Caroline. Pretty soon it was "Caroline wants to be with me all the time. Look at what a good daughter she is to me—more of a real daughter than my own." I wonder if she had any idea how deeply those words hurt and how guilty they made me feel. I found myself spending much more time with her than I wanted to, trying to fix her problems, just to avoid being compared to Caroline.

The person we're being compared to seems to get all the love and approval we want for ourselves, and it's only natural that we react competitively and try to get into that position ourselves. For Leigh, there was no end to the comparisons, and no way to measure up.

Dangerous Pressure

In the workplace, negative comparisons create an atmosphere like that in an unhealthy family, fostering envy and competitiveness. We may find ourselves trying to measure up to the impossible standards set by a paternalistic boss who, by pitting workers against each other, creates "sibling rivalries" among co-workers.

When my client Kim first came to see me, she was under intense pressure from a boss who used negative comparisons to "motivate" her. Kim, who is in her mid-30s, had the misfortune of being hired to replace Miranda, a legendary magazine editor who was about to retire.

I'm pretty efficient, and I have a lot of good ideas. I work well with writers, and I used to love what I do. But my boss is pushing me harder than anyone else on the staff, and he's constantly comparing me to Miranda. It's like I can never do enough. If I make four assignments in a week, my boss, Ken, says something like "That's great. Miranda used to do it in a slow week. Her record was eight or nine." If I say I need to leave on time one night instead of putting in my usual 10 or 11 hours, he'll talk about how the work ethic will be dead without Miranda. She's legendary for practically living at the office.

I believe that Miranda was brilliant—but she drinks like a fish, she has no family and lives for her work. The thing is, I'm trying to compete with her—and I do *have a life. I need to spend time with my kids and my husband, and I do work at my temple, that's really important to me. But Ken's always after me to do more, and sometimes when he says I could be the next Miranda if I take on just one more project, I fall for it. He's got me coming and going. If I don't do what he wants, he says I don't measure up to her. And then he'll also say that I have the talent to be a star like her—if I do the extra work he wants. He says I shouldn't think of it as extra work, I should think of it as job security.*

My family's going crazy because I'm not there, I'm getting exhausted, I'm starting to get pains in my arms and my neck from being at the computer so much—and worst of all, I'm questioning my own competence. But I feel like I have to measure up to the Miranda standard, like I won't be good enough until I do.

When we think about pressure in the workplace, we tend to pay the most attention to overt pressure—like getting fired. But the workplace can reactivate many of the same feelings and relationships that we have within our families, and the same dynamics come into play. Such issues as competition, envy, sibling rivalry and wanting to please authority figures can also drive us to our limits and beyond. The danger is that in trying to measure up to a tough standard set by someone with vastly different needs, talents and circumstances, we can find ourselves sacrificing our families, our outside interests and even our health to our jobs.

We're often quite definite, at first, about what we need and why we're resisting an emotional blackmailer. But the blackmailers' tools erode whatever clarity we have and persuade us that we really don't know what we want. Using these behavioral strategies, the blackmailer can almost always get our compliance—hardly surprising, considering that a person who resists is likely to be spun around, criticized, ganged up on or found wanting. Yes, this sounds dire, but it's all learned behavior that we have helped teach. And as we'll see, just as we've effectively handed blackmailers their tools, we can also take them away or render them useless.

5

The Inner World of the Blackmailer

Emotional blackmailers hate to lose. They take the old adage "It doesn't matter if you win or lose, it's how you play the game" and turn it on its head to read "It doesn't matter how you play the game as long as you do not lose." To an emotional blackmailer, keeping your trust doesn't count, respecting your feelings doesn't count, being fair doesn't count. The ground rules that allow for healthy give-and-take go out the window. In the midst of what we thought was a solid relationship it's as though someone yelled "Everyone for himself!" and the other person jumped to take advantage of us while our guard was down.

Why is winning so important to blackmailers, we ask ourselves. Why are they doing this to us? Why do they need to get their way so badly that they'll punish us if they don't?

THE FRUSTRATION CONNECTION

As we struggle to understand what turns the people close to us into emotional bullies, we need to return to the place the blackmail began—the moment when the blackmailer wanted something from us and our words or actions said no.

There's nothing wrong with wanting. It's fine to want, to ask and to try to figure out how to get what you want. It's OK to plead or reason and even to beg and whine a little—as long as we let a firm *no* mean *no*. Accepting no as an answer may not be easy, and the other person may get upset or angry for a time, but if the relationship is working, the storm will pass and we'll try to negotiate a resolution or compromise.

As we've seen throughout this book, however, that's just the opposite of what happens with a blackmailer. Frustration becomes the trigger not for negotiation but for pressure and threats. Blackmailers cannot tolerate frustration.

It's hard to understand why they make such a big deal of this. After all, many of us have faced plenty of disappointments without becoming bullies to feel better. We accept disappointment as a temporary setback, and we go on. But in the psyche of blackmailers, frustration symbolizes something far beyond being blocked or disappointed, and when they encounter it they can't just regroup or shift gears. To the blackmailer, frustration is connected to deep, resonant fears of loss and deprivation, and they experience it as a warning that unless they take immediate action they'll face intolerable consequences.

FROM FRUSTRATION TO DEPRIVATION

On the surface, blackmailers appear to be just like everybody else, and they're often highly effective in many areas of their lives. But in many ways, the inner world of the blackmailer resembles Depression-era America—that terrible time when many hard-working people saw their lives unravel and their security replaced with sweeping deprivation. If you know people who lived through that era, you've probably noticed how many of them still hoard and pinch every penny, bracing for another shock, another loss, as they try to reassure themselves that they'll never have to face pain like that again.

Emotional blackmailers—no matter what their style or preferred tools—operate from a similar sort of deprivation mentality, but we may not glimpse it until something happens to shake their sense of stability and stimulate their fear of deprivation. Just as some people can interpret a headache as a sure sign of a brain tumor, blackmailers see resistance as a symptom of something much graver. Even mild frustration is viewed as potentially catastrophic, and they believe that unless they respond to it aggressively, the world—or you—will keep them from getting something they vitally need. A deprivation tape starts to play in their heads:

- This isn't going to work out.
- I never get what I want.
- I don't trust other people to care about what I want.
- I don't have what it takes to get what I need.
- I don't know if I can stand it if I lose something I want.
- Nobody cares about me as much as I care about them.
- I always lose anyone I care about.

With these thoughts cycling through their minds in an endless loop, blackmailers believe they haven't got a chance of prevailing—unless they play hardball. That belief is the common denominator underlying all emotional blackmail.

DEPRIVATION AND DEPENDENCY

For some blackmailers, these convictions are rooted in a lengthy history of feeling anxious and insecure, and if we look far enough back in their lives, we can often find important connections between incidents in their childhoods and some of their adult fears of deprivation.

Allen, the businessman whose wife used blackmail to keep him from making plans that didn't include her, began to see some of the underpinnings of her behavior when Jo became especially moody around the anniversary of her father's death.

I asked her if I could do anything to cheer her up, and she pulled out some pictures I'd never seen from her junior high graduation. Her father had died two nights before the photos were taken, and she looked like a scared little girl trying to put on a smiling face. It turned out that she had to handle everything after his death—calling the relatives, making funeral arrangements, even getting herself ready for the big ceremony at school because she was giving a speech that her father had helped her write. She had to keep herself together because everyone else in her family fell apart. I asked her mom about this recently, and she said that Jo didn't even cry very much. She just kind of retreated into her room.

Jo told me that she'd never loved anyone as much as she loved her father, and then suddenly he was gone. I think she's always been afraid I'll leave, too, and her holding on is a way to keep me.

Emotional blackmail is the only way Jo knows to cope with a world she doesn't trust, a world she is certain will rob her of what she loves. People who have faced major deprivation and loss in childhood often become needy and overly dependent as adults to keep from feeling rejected, abandoned or ignored.

Jo had done well at school and felt cherished by her father, but none of that was enough to protect her. The profound sense of childhood helplessness stayed with her, and as an adult she tried to make up for it by developing elaborate strategies to keep herself from experiencing pain of that magnitude again. She learned to cling fiercely to friends and lovers, but she'd never found an *appropriate* way to express her fear that, no matter what she did, they would be taken from her.

When she married Allen, her fears escalated. Instead of being able to enjoy their relationship, she became frightened any time his plans didn't include her. She convinced herself that if she could keep him with her all the time, she could not only keep from losing him but also regain some of the security she'd lost when her father died. She had a core belief common to many emotional blackmailers: I don't trust that I'm going to get what I need, so I have to give myself every advantage. That justified all the clinging and all of the blackmail.

A MIX OF CAUSES

The primary roots of Jo's deprivation were fairly easy to trace, but bear in mind that human behavior is complex, shaped by a variety of physiological and psychological factors. Often it defies any single explanation. We're born with particular temperaments and genetic predispositions—our own unique circuitry—and these elements interact with how we're treated and what we learn about ourselves and our relationships to others, shaping our inner and outer lives.

Eve, whose artist boyfriend, Elliot, was hypersensitive to frustration and frequently threatened to harm himself when he felt endangered, once told me about a conversation she'd had with Elliot's sister.

She laughed when I asked her if she knew why he was always throwing tantrums. She said he'd been that way since he was born. Even when he was a baby, if the bottle wasn't right in his mouth or he was wet for two seconds, he'd yell the house down. When he was a little older, he was a terror with his tantrums. She said it's just how he's wired—he was the most demanding, needy child she's ever seen.

That child grew up to be a demanding, needy adult who continued to throw tantrums to get what he wanted. A large part of Elliot's basic temperament was present in infancy, including his low frustration threshold.

Complementing or reinforcing such genetic factors, of course, are powerful messages from our caretakers and society about who we are and how we are supposed to behave. Defining experiences in childhood, adolescence and even our adult lives create potent beliefs and feelings that often resurface, especially in conflict or when we're under stress. We revert to those old patterns because they're familiar, and even though they may cause us pain, they provide reassuring structure and predictability. We also believe that even if the old behavior didn't work for us before, when we repeat it this time we'll get it right.

Like Jo, many blackmailers have a fantasy that the helplessness and inadequacy they felt as children will vanish, and now, as adults, they'll magically be able to "fix" a bad situation or an unhappy parent or guarantee the security they long for. They believe they can compensate for some of the frustrations of the past by changing the current reality.

WHEN A CRISIS IS THE CATALYST

The inability to tolerate frustration can also be a response to fairly recent uncertainties and stresses. The potential for blackmail rises dramatically during such crises as a separation or divorce, loss of a job, illness and retirement, which undermine blackmailers' sense of themselves as valuable people. Most of the time they aren't even consciously aware of their newly activated fears. All they can see is what they want right now and how they'll get it.

For Stephanie, the precipitating crisis was her husband's confession of his brief affair. Bob was working hard to get the marriage back on stable ground and attending therapy regularly, but Stephanie remained adamant about her right to keep Bob in line with healthy doses of emotional blackmail. After a year of her anger and retaliation, Bob was almost ready to give up. I told him it would be a good idea for her to come in for a joint consultation, and she agreed.

You of all people should understand. I've read all your books and you talk a lot about how you can't let people roll over you and how you have to confront them and set limits. I have every right to be mad, and Bob deserves some payback for what he did.

I told Stephanie that she was indeed entitled to feel angry, hurt, betrayed and shocked, and I didn't want to discount her pain in any way. But, I told her, there is a big difference between confrontation and emotional blackmail. While she might be getting satisfaction from playing the avenging, wronged wife and punishing Bob, her marriage was going down the tubes.

Stephanie became noticeably less defensive as the session went on, and as she tearfully described how she'd felt when she found out about Bob's affair, another layer emerged, shedding considerable light on why it was so difficult for her to let go of her need for revenge.

This wasn't the first time I gave my heart and soul to a man who let me down, and Bob knew that. How could he even think of seeing someone else when he knew how devastated I was when my first husband cheated on me? It nearly killed me. What am I supposed to do now? How can I ever trust him again? I have never felt so unattractive, so humiliated, so . . . so inadequate in my life!

Stephanie was not only coping with what Bob had done, which would have been difficult enough, but also with the pain of the experience with her first husband. Drained of trust in Bob and of confidence in herself, she struck back through punishing emotional blackmail, the one way she felt she could regain control over the emotional chaos inside her.

While issues from Stephanie's childhood had probably played a part in her reactions, we focused on the parallels in her adult life. When Stephanie saw how much the pain she'd brought from her former marriage was jeopardizing what could once again be a good relationship with Bob, she agreed to see a colleague of mine. She and Bob are both working hard, and they have been able to use this crisis as a catalyst for opening up new areas of communication and exploration. I think they'll make it.

A CHARMED LIFE

Some of the most puzzling blackmailers are the ones who seem to have it all and want more. It seems incongruous to suggest that they're motivated by deprivation because they seem to have had so few encounters with it. But often, people who have been overprotected and indulged have had little opportunity to develop confidence in their ability to handle any kind of loss. At the first hint that they might be deprived, they panic, and shore themselves up with blackmail.

This was definitely the case with Maria's physician husband, Jay. I knew from my work with Maria that Jay was a man to whom everything had come easily. He'd flown through medical school, established himself as a wunderkind by pioneering a number of surgical procedures and moved easily in the most prestigious social circles. The word that came to my mind was *entitlement.*

His childhood was incredible. No abuse, no trauma, nothing but adoration. His father was poor and the first one in his family to go to college. That man was amazing. He got through medical school with a combination of hard work, chutzpah and about two hours' sleep a night. He had a part-time job as a waiter to earn enough money to take Jay's mother out. He told me one of the things he'd promised himself was that Jay should never have to go through what he did. Jay was a golden child, no question. When he decided to become a doctor his parents spared no expense to see that he got chemistry sets and science camps. No part-time jobs for him. He got the world on a platter—along with tennis lessons, cashmere sports coats and, of course, lots of girls.

Jay's life had been more than privileged—it had been unreal. In trying to make sure that deprivation had no place in Jay's life, his father had done little to prepare him for disappointments or setbacks.

There are two down sides to this charmed existence. People who grow up in such a safe harbor develop unrealistic expectations, believing that whatever they want will fall into their laps. Even more damaging, they are deprived of the chance to develop the essential skills we all need for handling frustration. With what appeared to be the best of motives and intentions, Jay's father had actually created a kind of emotional cripple.

When Maria challenged Jay's belief that he was entitled to have it all—career, family, wife and mistress—she was the first person who mattered to him who threatened to take something important away from him. Jay was panic-stricken. Someone had changed the rules, and emotional blackmail was his means of regaining his accustomed position at the top of the heap.

INTIMATE STRANGERS

When Jay enlisted his parents to pressure Maria to stay with him, she couldn't believe what they were saying to her:

My God—what am I dealing with here? People I loved and respected suddenly show themselves to be without any moral or ethical convictions whatsoever. Is keeping up appearances more important than feelings and basic human decency?

Maria watched Jay change from the charmer who'd swept her off her feet to a devious, manipulative stranger. When people close to us resort to emotional blackmail, we're struck by how their personalities seem to shift, a process that may be gradual or surprisingly fast. Much of the pain and confusion of emotional blackmail, in fact, arises from seeing people we care about and who we believe care about us become people who need to get their way so much that they are willing to ride roughshod over our feelings.

Liz had a feeling of disorientation when Michael told her how brutally he'd punish her if she kept talking about leaving him.

He's actually said: "The money you'll have left when I get through with you won't buy dog food. And kiss the kids good-bye. I'm thinking of taking them to Canada so they won't have to listen to your lies about me." And this is a man I've made love with and been naked with and bared my soul to. Who the hell is he?

Blaming, threatening, negative comparisons and other tools we've seen are obviously not what drew us into our close relationships, and they're not what's kept us there. These are people who share our lives, our work, our feelings and our secrets. Yet when emotional blackmail comes in, we quickly encounter some unappealing aspects of their personality—self-centeredness, overreaction, an insistence on short-term gains even if they result in long-term losses, and a need to win, no matter what.

IT'S ALL ABOUT THEM

All the blackmailers we've seen are focused almost totally on *their* needs, *their* desires; they don't seem to be the least bit interested in *our* needs or how their pressure is affecting us.

Blackmailers can be like steamrollers when we don't satisfy them, becoming ruthless in their single-minded pursuit of their goals. It's a strange kind of love that is so blind to the target's feelings.

My nominee for the Narcissist Hall of Fame is Patty's husband, Joe, who took to his bed with a contemporary version of the vapors when she told him they couldn't afford a new computer. In a recent incident, his self-absorption was unparalleled:

Joe makes good money, but he spends it faster than either one of us can make it, so we're usually behind on everything. Last week the bills were really piled up, and he asked me to call my aunt and ask for a loan. My aunt is pretty well off, but she'd just had surgery for breast cancer. I told him there was no way I could bother her, and I couldn't believe how he started to pressure me. "Here's her number at the hospital," he said. "You don't even have to look it up. Call her now—it's no big deal. She's not in any pain and you've always been her favorite. Why can't you do this one little thing for me?"

Breast cancer? Hospital? Surgery? No problem for this blackmailer. After all, he wanted something. NOW. And at the moment of urgency, no one else existed on the face of the earth.

Often blackmailers' self-centeredness springs from a belief that the supply of attention and affection available to them is finite—and shrinking fast. Elliot is so self-centered that even when his girlfriend wants to take some classes to provide a backup career for herself, it's about him. In his mind, giving to Eve robs him of security. What if he needs something while she's gone? What if he gets bored or lonely? Who will take care of him? The universe revolves around him, just as it did when he was little. Once again, he's the tyrannical five-year-old demanding total attention and satisfaction from the person he's dependent on, and always wanting more.

MAKING MOUNTAINS OUT OF MOLEHILLS

Emotional blackmailers often behave as though each disagreement is the make-or-break factor in the relationship. They feel such intense disappointment and frustration when they encounter resistance that they blow up even minor discord and allow it to color the whole relationship. Why would anyone make such a big deal out of your not wanting to have dinner with their parents, or attack you because you wanted to take a class or a fishing trip, or weren't enthusiastic about their plans? The ferocity of their desire only makes sense when you realize that they are not reacting to the current situation but rather to what that situation symbolizes to them from the past.

It became clear from some things Eve told me about Elliot's background that he believed that a man can't get what he needs from an independent woman.

I remember him telling me about his dad and how much he groused about being neglected. Elliot's mother must have been a real pioneer in the business world. She ran a small children's-clothing company, which was great—except for her husband, who hated it. Elliot said the thing he remembers most was his mother being gone a lot. She was loving when she was there, but then suddenly she'd be off on a business trip and he missed her terribly. His father was angry at her much of the time and was constantly saying things like, "These goddamn women—they can't do enough for you when they need you, but they forget you exist if they start making it on their own." I guess if you hear that enough it really sinks in.

The message Elliot came away with was unambiguous: Women won't be loving partners unless you make sure they're with you all the time. He would probably deny holding that kind of belief, but his overreaction to Eve revealed that old demons were being stirred up. For Elliot, any sign of independence in a woman was threatening. Eve became a stand-in for his mother, the woman he was emotionally dependent on and fused with. She, too, would abandon him, as he perceived his mother had abandoned his father—and him—by being gone so much. Each time Eve tried to walk out the door, Elliot began to relive his old sense of deprivation.

As with all overreaction, a lot of noise and emotion are vented, but the real, underlying feelings are rarely expressed. Elliot is desperate for intimacy, but the barrage he directs at Eve almost guarantees that he won't get it. Let's look at what was expressed, and unexpressed, when Eve suggested that he might consider getting professional help for his anxiety.

What Elliot says: "You're gonna go out and do what you want and I'm gonna be alone—why should I live? You don't care a bit about me."

What Elliot means: "I'm scared because you're changing. At the beginning, I was enough for you, but not anymore. If you go to school, I'm scared you'll get a career of your own, and you won't have time for me. I'm scared you'll meet someone else. I'm scared you'll get too independent. You won't need me and you'll leave me."

But this kind of communication was not in Elliot's repertoire. If it were, he wouldn't need to resort to emotional blackmail. Perhaps, like many men, he was ashamed of his neediness and fears. The only option he saw for getting what he wanted was to rant and rave—overreacting to the slightest indication of Eve's wish to improve herself.

Feelings from Another Time

Roger, the screenwriter, was astounded at the furious assault Alice unleashed against him when he didn't jump at her suggestion that they have a baby. When he appeared uncertain about what he wanted, she had a classic overreaction.

You never really cared about me. How can you call this love when you don't want to take any steps that will bring us closer. I don't trust you anymore. I'm not even sure I love you anymore! You have some pretty serious problems, and you really need help!

However, one night at their AA meeting, Roger was able to gain important insights into the fears behind Alice's desperate push to get a long-term commitment from him. She told the group:

I don't believe in anything but right now. *I dive into the present and hold on tight. My father was a compulsive gambler. He was really glamorous in my eyes. But what it really means is that one day you're rich and the next day you're shopping at the Salvation Army and you can't answer the phone because it's a collection agency. Everything I had as a kid was up for grabs—baby-sitting money, gifts from people, anything you could pawn. Even my father came and went. Sometimes he would be gone for weeks. Is it so wrong to want some security, some commitment? That feels like good values to me. What's wrong with love?*

Alice had spent years dreading that what she had would suddenly disappear, so it was no surprise that she wanted guarantees for the future. But like most blackmailers, she used a heavy-handed style of attempting to get around her target's resistance.

Her overreactiveness, which took the form of a scathing verbal attack on Roger, came from a place inside her that was full of hungers and fears. No matter how much she bound Roger to her, he couldn't fill that place for her, even if he wanted to.

Through her work in AA, Alice was able to see how she was trying to handcuff Roger and to realize that she would have a difficult time with any relationship until she did some work on herself. She has since been able to significantly reduce the pressure on Roger and to give the relationship time to evolve naturally.

WINNING THE BATTLE, LOSING THE WAR

Blackmailers frequently win with tactics that create an insurmountable rift in the relationship. Yet the short-term victory often appears to be enough of a triumph—as if there were no future to consider.

Most blackmailers operate from an I-want-what-I-want-when-I-want-it mind-set. They seem to have a childlike inability to connect behavior to consequences, and they don't appear to give any thought to what they will be left with once they've gotten the target's compliance.

It's hard to believe that Michael, Elliot, Alice, Jay, Stephanie or any of the blackmailers we've seen so far could think they would have anything worthwhile left if their targets gave in to their threats and pressure. What kind of relationship does Josh's father expect to have with his son if Josh provides the victory his father is demanding by giving up his lover? Margaret, whose husband, Cal, pressured her to take part in group sex, gave in to her husband's emotional blackmail, but it was the death knell for the marriage.

Liz bought herself time to regroup by appearing to give in to Michael's threats. As Liz said,

I called my lawyer and told him to hold off on doing anything. I keep hoping now that Michael's calmed down he'll at least start a rational dialogue with me. He's really being sweet because he thinks he's got me where he wants me and that ultimately I'll come around and kiss and make up. The truth is, I'm just going through the motions. I'm living with a man I don't even like anymore, let alone love.

Any logic or ability to see the consequences of their actions is obscured by the urgency blackmailers feel to hold on to what they have. They're in a FOG of their own that makes them oblivious to how much they alienate other people with their bullying. All that matters is finding immediate relief from their deprivation fears—whatever that relief costs.

THE PAYOFFS OF PUNISHING

When we look at how intensely blackmailers fear deprivation, a more complete picture begins to take shape and it's easier to fathom why they act as they do. But one question continues to nag most of the people I talk to about this subject: Why do they need to punish us? "OK," they say. "I see why they nag or pressure us or even make threats, but why on earth do they need to hurt us if we don't give them their way?"

Many times it *does* seem as though the goal of emotional blackmail is not only to make the blackmailer feel good but to make the target feel bad. Blackmailers demand—and they demean. In their attempts to show the rightness of what they want, they cast aspersions on our character and question our motives. Even when their threats of punishment are directed more against themselves than us, they target us by drowning us in guilt.

One obvious reason is the difference between what blackmailers *tell* themselves about what they're doing and why they're doing it—the "spin" we saw in the last chapter—and the actual effect of their behavior on us. Punishers don't see themselves as punishing, but rather as maintaining order or keeping a firm hand on things or doing "what's right" or letting us know they can't be pushed around. They see themselves as strong and in charge. If their behavior hurts us, so be it. The end justifies the means.

And as we've seen, many punishers see themselves as victims. In fact, the more abusive the blackmailers, the more they twist reality. Their extreme sensitivity and self-centeredness magnify the hurts they feel and help them justify retaliating against us for what they see as deliberate attempts to thwart them.

Punishment also allows blackmailers to take an active, aggressive stance that makes them feel powerful and invulnerable. This is an extremely effective way for them to calm any perceived threat of deprivation and head it off at the pass. After all, if someone is yelling, threatening, slamming doors or refusing to speak to you, there's not much spare time to deal with feelings.

It's a truism that what we don't talk out, we act out. If punishers were to allow themselves a few moments of introspection, they would probably be revolted by the fears and vulnerabilities they would discover. It's one of the most fascinating paradoxes of human behavior that angry, punitive people are really very frightened, but they rarely confront or diminish those fears. Instead they lash out at others when they are frustrated to prove how strong they are. They create so much unhappiness with their behavior that they often cause people to leave them, thereby ensuring that the thing they fear the most will happen.

CUTTING THEIR LOSSES

The most punitive blackmailers are often those who have lost or fear losing someone important to them because that person is withdrawing emotionally or because of a separation, divorce or major rift in the relationship.

Remember Sherry and Charles, the married man she worked for who threatened to fire her if she ended their affair?

One minute I'm the most beautiful, exciting, interesting woman in the world, but as soon as I tell him I feel like I'm on a dead-end street and I have to get off so I can move on with my life, I'm a cold-hearted witch who doesn't care about all the stress he's under and how much he's trying to work things out. Now he's telling me that all he does is give, give, give, and all I do is take, take, take—which is a total 180 on what's been happening. Oh—and suddenly he's finding fault with everything I do at work. If he's trying to make my life miserable, he's doing a good job of it. How could he turn on me like this?

Faced with the loss of his young lover, and seeing that his threats weren't working, Charles did something to ease his pain—he devalued her. If he could make her less desirable, less valuable, then he wasn't losing so much, and his deprivation would be considerably alleviated. After all, damaged goods are easier to part with. He could also justify firing her by devaluing her work. Double devaluation, double punishment.

Devaluation is a common tactic for angry blackmailers. It softens the sting of confrontation and enables them to downplay their feelings of loss. But as they do this, they give their targets confusing double messages. It's as if they were saying "You're no good, but I'll do everything in my power to hold on to you"—a further illustration of how desperate they're feeling.

Even though the last thing in the world they want is to end a relationship, they will often initiate the breakup on their own if they sense that their partner is serious about leaving. Their aggressive stance will let them stay in control, the old face-saving strategy that goes "Let me quit before I get fired."

TEACHING US A LESSON

Like parents who believe that punishment will mold a child's character, blackmailers may convince themselves that they're helping us with their punishments. Instead of feeling guilt or remorse about hurting someone they care so much about, they can actually feel pride. They're making better people of us, they reason.

Alex, the tantalizer we met in Chapter 2, believed that by promising his girlfriend Julie help and contacts but withholding his offers unless she "measured up," he was doing her a great favor.

He told me that sending my son off to my ex's would be the best thing for me. Everything was in terms of "You're holding yourself back" and "I just want to see you reach your potential," when what he really wanted was to have me to himself without my little boy around. Right, he was really generous.

Insults and infantilizing are similarly explained away with the "It's for your own good" rationale. There's generally much less malice here than you might think. Most blackmailers believe they really are teaching us valuable lessons. Charles was absolutely in earnest when he told Sherry, "You need to learn about loyalty. It's the most important thing in this business."

Lynn and Jeff, who blackmailed each other, also thought they were trying to improve each other's character. "She's got to learn that she can't treat people like that," Jeff told me after one of their fights, sincerely believing that he was teaching Lynn to be "less of a shrew." And Lynn saw her actions as training, too. "Maybe if I humiliate him enough, he'll get off his ass and get a part-time job," she said. "Sometimes it takes a kick in the pants to get someone moving."

It's fairly apparent, especially to the person on the receiving end, that punishment doesn't produce the results the blackmailer believes it will, yet there are attractive payoffs to clinging to this erroneous idea of punishment as training. Blackmailers can live with almost anything if they can make their targets seem like dunces. In this way they can avoid any introspection or hint that something in *them* is driving off the love or connection they so desperately want.

OLD BATTLES, NEW VICTIMS

As we've seen, sometimes current stresses in the blackmailer's life reopen old wounds, and the blackmailer strikes out at a target who has become a stand-in for a figure from the past. When this happens, a blackmailer's punishment can be overblown, and may feel completely uncalled-for.

Michael, perhaps the most blatant punisher we've met, began to seem like a monster to Liz, who felt almost shell-shocked in the face of his tirades. When I asked her why she thought he was being so mean to her, she was silent for a moment, then responded:

You know, when I really think about it, I think that Michael was like a powder keg that was ready to blow. He's been working hard since he was about 14 in his family's business. They sell office equipment, and they've been pretty successful. Michael never got to be a kid, really. He was great at sports—he's still pretty athletic—but Mom and Pop never let him play. He had to do inventory or sweep the store or mind the register.

When we were first dating, we took a trip to Chicago and he knew everything about the buildings. And he told me his big dream was to go study architecture. But they never let him go, and he let it drop. He's so responsible. I know he's mad at them, though he's never said a word and never will. But I don't think that means he has to take it all out on me.

I told Liz that she was right, there's no excuse for the kind of verbal attacks and threats Michael had leveled at her. But it was important for her to understand that Michael's criticisms of her and his anger at her supposed flaws had nothing to do with her, even though it was difficult not to take them personally. When Liz got fed up and threatened to leave him, the punishments reached a fever pitch. His fear of being without her had ignited the stored-up frustrations Liz had seen.

If Michael had had the ability to express what he was feeling, he might have told her: "Please don't take my dreams away again. I've been disappointed, hurt and deprived since I was a teenager, and I never really got what I wanted. No one cared about me, and that really hurts. How could my parents have destroyed everything I loved and forced me into this job I hate? And now you want to leave? I can't stand to see this happen again. How much disappointment do you think I can handle?"

That's a speech full of emotions that rightfully deserve to be directed at his parents, but having been controlled by them all his life, he's never felt safe or strong enough to make it. All the sadness and anger that Michael had stuffed didn't disappear. It blew up into his life. He confused Liz, whom he loved, with the parents he had grown to resent.

MAINTAINING A STRONG CONNECTION

Odd as it seems, punishment keeps a blackmailer in a strong emotional connection with you. In creating a highly charged atmosphere, blackmailers know they are activating the target's feelings for them, and even if the feelings are negative, they've created a tight bond. You may resent or even hate the blackmailer, but as long as your focus is on them, they haven't been abandoned or discarded with indifference. Punishment keeps a lot of passion and heat in a fractured relationship.

Allen's ex-wife, Beverly, continued to punish him in the most painful way possible—by using their children as a weapon. Allen and Beverly had had an acrimonious divorce. Even though their marriage had been a source of unhappiness and stress for both of them, it was a divorce that he wanted and she didn't. And Beverly fought him tooth and nail. They had made a few attempts to reconcile, and even tried counseling, but to no avail.

She knew what those kids mean to me. I don't think a lot of people really understand what it's like for a man not to be able to be with his kids every day, especially while they're growing up. I had to divorce Beverly, but I didn't want to divorce my kids. At first she threatened that if I left her I'd never see them again. She'd move out of the state, maybe even out of the country. I was panicked. I couldn't think straight. I've known women who've done that—hell, I've known men *who've done it to their ex*-wives.

Everyone finally settled down, and ultimately Allen was given liberal visitation rights. He and Beverly managed to be civil to each other, and she honored the court order. But his remarriage to Jo started the blackmail up again.

Now that I've got someone I really care about in my life, she can't stand it. I guess as long as I was single she figured there was still a chance. I know she's still pretty bitter. So she's getting back at me again through the kids. If I'm ten minutes late picking them up, she'll have taken them out somewhere. It's almost an hour's drive to where they live, and I can't always get there on the dot. Last week I had to wait almost an hour and a half. When she pulled into the driveway she said, "I'm not going to wait around for you. How did I know you'd even show up?" She expects me to take it and not complain. But if I need to reschedule something, she goes ballistic. And if the child-support payments are a day late, she's on the phone threatening to take me back to court to get my visitation reduced. Christ, we're talking more now than we did when we were married!

Allen's ex clearly hasn't let go of him or the marriage. And like many divorced blackmailers, both women and men, she's using the most powerful weapon in her arsenal—their children—to keep an emotional connection with him. Allen and Beverly are legally divorced, but the psychological divorce is still pending.

The use of children as a weapon against the noncustodial parent is one of the oldest and cruelest forms of emotional blackmail. There are no higher stakes. It's especially effective because of the intensity of the emotions involved. It keeps people who once cared about each other locked in a terrible battle in which everyone loses.

IT'S NOT ABOUT YOU

The most important thing to take away from our tour of the blackmailer's psyche is that emotional blackmail *sounds* like it's all about you and *feels* like it's all about you, but for the most part *it's not about you at all.* Instead, it flows from and tries to stabilize some fairly insecure places inside the blackmailer. Much of the blaming, spin and self-righteousness that have made us feel so bad about ourselves—often bad enough to give in to the blackmailer's pressure—is not valid. It's fear-based. Anxiety-based. Insecurity-based. And those fears, anxieties and insecurities reside inside the blackmailer. Many times emotional blackmail has more to do with the past than with the present, and it's more concerned with filling the blackmailer's needs than with anything the blackmailer says we did or didn't do.

That's not to say we don't play a pivotal role in the blackmail process. After all, it couldn't happen without our compliance. Now it's time examine the factors inside *us* that allow blackmail to take hold.

6

It Takes Two

Blackmail takes two. It's a duet, not a solo performance, and it cannot work without the target's active participation.

I know that often that's not how it feels, and I know how natural it is to become defensive about our own behavior. It's always more comfortable to concentrate on what the other person is doing rather than to acknowledge what we bring to the table. But in order to break up the blackmail partnership, you'll have to turn your attention inward and look at the elements that have led you, often unknowingly, to participate in emotional blackmail.

Please remember that when I talk about participating in blackmail, I'm not suggesting that you provoke or cause it, but rather that you give it permission to occur. You may not even realize that someone's demands are unreasonable. It may seem that you're just being a good wife or a good employee or a good son or daughter—accepting another person's preferences almost unquestioningly because you've been taught that's what you're supposed to do.

Or you may be well aware of the blackmail but feel as though you can't resist it, because the blackmailer's pressure sets off almost programmed responses in you, and you're reacting automatically or impulsively. Keep in mind that not everyone responds to blackmail attempts with compliance. If you do, I want to help you get a handle on how and why. As a start, please think about and answer the following questions.

When faced with blackmailers' pressure do you:

- Constantly berate yourself for giving in to their demands?
- Often feel frustrated and resentful?
- Feel guilty and believe that you're a bad person if you don't give in?
- Fear that the relationship will fall apart if you don't give in?
- Become the only one they turn to in a crisis, even though there are others who could help?
- Believe the obligation you have to them is greater than the one to yourself?

If you answered yes to even one of these questions, your responses to pressure are helping to create an ideal climate for blackmail.

HOT BUTTONS

Why is it that some people, no matter how smart or together they are, seem to be so vulnerable to emotional blackmail, while others can brush it off? The answer lies in our hot buttons, the sensitive bundles of emotional nerves that form inside all of us. Each hot button is like a power cell charged with our unfinished psychological business—stored-up resentments, guilt, insecurities and vulnerabilities. These are our soft spots, shaped by our basic temperament and sensitivities along with our experiences from the time we were little. Each spot, if probed, would reveal vivid layers of our personal history—how we were treated, what self-image we carry inside and how we've been marked by impressions from our past.

The feelings and memories stored in our hot buttons can be searing, and when events in our current lives remind us of something we've kept buried inside, they can bring up reactions that override thought or logic, tapping pure emotion that has been stored and gaining power for a long time.

We may not always remember the incidents that led to the formation of our hot buttons, and when it comes to the complexities of why we do the things we do, cause and effect can be elusive. But if you've ever wondered where "stuffed" feelings and experiences go, your hot buttons are a good place to look.

Providing a Map for the Blackmailer

Over the years our emotional landscape may become studded with these hot buttons, and many of us design portions of our lives to navigate around them. In fact, the most common strategy for people attempting to deal with these hot areas can be summed up in four words: avoid at all costs. We may not realize what we're doing, but as we take this path of avoidance, we reveal ourselves more clearly than we know. As we tiptoe around our hot buttons, we're practically drawing a map of where they are, which those who know us well can't fail to see.

We all know what the people around us are sensitive to; it's no secret when a friend is afraid of anger or tends to soak up blame. But for the most part, we're fairly compassionate and don't use that knowledge to our own ends. When our blackmailers are feeling safe, they don't either. But faced with resistance, their fears of deprivation flare up. They drop the compassion and use every bit of information they have about us to ensure that they prevail.

THE TRAITS THAT MAKE US VULNERABLE TO BLACKMAIL

To insulate ourselves against having our hot buttons activated, we develop a number of specific personality traits. They're so much a part of us that it may not be apparent at first that they're defenses against what we fear. But as we look at them more closely, you'll see that all are deeply related to our hot buttons. Ironically, it is these "protective" qualities that open us up to emotional blackmail. They are:

- An excessive need for approval
- An intense fear of anger
- A need for peace at any price
- A tendency to take too much responsibility for other people's lives
- A high level of self-doubt

None of these traits is harmful—in moderation. In fact, some are viewed as positive, and even rewarded, when they're not taken to extremes. But when they control us and go to war with the intelligent, confident, assertive and thoughtful parts of ourselves, we're setting ourselves up for major manipulation.

As we examine these traits and the behavioral styles they give rise to, notice how much of the target's behavior is actually a response to feelings from the past. Notice, too, how often targets are betrayed by the very responses they believe will protect them.

THE APPROVAL JUNKIE

It's perfectly normal to want the people we care about to approve of us—we all want their goodwill. But when we *must have* it, and it becomes a drug we cannot do without, we aim a spotlight at a hot button that a blackmailer can easily zero in on.

In the introduction, I talked about my client Sarah, who constantly had to prove herself to her boyfriend Frank. Each time she passed one of his tests, she basked in the glow of his approval. But each time she protested, he took it away, and she felt miserable. So she tried to maintain an uninterrupted flow of approval by giving in to his pressure—even when what he wanted went against the grain for her.

I can't stand it when he's upset with me. When I said that painting the cabin was not how I had expected to spend the weekend, he just shook his head and went out on the porch. I followed him out there, and he said he couldn't believe how spoiled and infantile I was being. I felt scared, and really shaky. So I went in and put on some old clothes and grabbed a brush. Then I got that great smile of his and a hug, and I could breathe again.

She'd gotten her "fix." There's nothing wrong with wanting approval, or asking for it. But approval junkies need a constant supply, and judge that they've failed if they can't get it. They believe they're not OK unless someone else says they are, and their sense of security depends almost entirely on outside validation. The approval junkie's motto is "If I'm not getting approval, I've done something wrong." Or worse yet, "If I'm not getting approval there's something wrong with me."

Sarah's description of how devastated she felt when Frank was upset with her indicated a driving need for approval and a terror of what would happen if she didn't regain it. This fear directly parallels that of a young child. In the child's fantasy, the consequences of losing approval are catastrophic. "I did something Daddy (or Mommy) didn't like. Daddy's upset with me. Daddy doesn't love me anymore. Maybe Daddy will get rid of me. I'll be alone and I'll die."

Sarah discovered that her tendency to link approval with survival came not so much from her parents but from her grandmother, who had taken care of her while they worked.

God, she was formidable! She lived in a downstairs apartment my parents had set up for her, and that's where I went every day after school. She was always critical—she told me I was too loud and that I was lazy. She said God didn't like lazy girls, and sometimes they got sent away. I don't think she meant to be mean, and I'm sure somebody told her something ridiculous like that when she was little. But it scared the hell out of me. She taught me a rhyme that I don't think I ever put out of my mind. "Good, better, best. / Never let it rest, / Until good is better / And better is best."

Sarah learned many lessons from her grandmother, whom she adored and with whom she spent many hours in her formative years. Some have worked well for her, others less well. She learned that if she behaved in a way that got her grandmother's approval, she was a good girl, and therefore safe. But she also learned that what she did was never quite good enough for her perfectionistic grandma, and that the elusive "best" was always just out of reach.

The feelings that Sarah describes with Frank—her compulsion to please and a fear of disapproval that seem almost to take on a life of their own—are familiar to all approval junkies, and a sure sign that someone has pressed a hot button.

When we're little, we need the approval of the powerful giants who care for us, and the ghosts of that need can haunt us long after we learn to take care of ourselves. In the home where Sarah had grown up, love was given and withdrawn based on how she "performed," and she developed a voracious need for someone else's OK. Frank tapped into that need when he withheld his "applause" and affection. Sarah knew intellectually that she couldn't please everyone 100 percent of the time—but she felt she had to try.

Sarah was primarily focused on Frank's approval. Maria, who was trying to withstand her husband's pressure to stay in their marriage even after she'd learned he was seeing other women, found herself dominated by a somewhat different concern: What will people think?

Divorce is not done in my family, or among the people I'm closest to. If that sounds old-fashioned, well, it is. I'm old-fashioned, and I'm proud of it. I can't stand the idea that I can't make this marriage work. And I can't begin to think about what would happen if I decided to leave Jay. What will people think? My life will fall apart at the seams. My parents will be upset with me, and his parents, and my kids, and my minister. They'll think I didn't have the guts to hang in there and fight for my marriage.

As Maria made her best case for remaining with Jay, it seemed that the weight of family tradition, history and community was pressing upon her, and she was sure that she had no choice but to stay. She was certain that she would be abandoning her principles if she considered a divorce. But as we worked together, Maria began to realize that the beliefs she held so strongly had been imposed on her—the ideas she defended so fiercely weren't even her own—and her definition of what makes a good family or a good marriage was far broader and deeper than "one that stays together no matter what."

For Maria, this discovery felt liberating, yet she was reluctant to explore or express the genuine beliefs she was uncovering in herself because she so needed to maintain the approval of her relatives, friends and church community. This woman, who held down a good job, ran a household, raised two terrific kids and managed an active church and social life, became a helpless child when she imagined the disapproval of people she cared about. As we dug for the roots of her approval hunger, a process that took many weeks, Maria remembered what she had always thought of as a "minor incident" that had taken place when she was a junior in high school.

I'd always been a Goody Two-shoes, but one day toward the end of the school year, my boyfriend, Danny, who was the love of my life, said no one would even know if we skipped last period and went to the beach. So we did, and I put it out of my mind. But a couple of days later, my father started asking me if I had anything I wanted to tell him. When I said I couldn't think of anything, he said he couldn't believe a daughter of his would lie to him, so he'd ask me again. Did I have anything to tell him?

My heart started pounding, but I couldn't confess. I just kept still. Then, in a very quiet tone of voice, Daddy told me that the school had called, and he knew what I'd done. I'd embarrassed him and our family, and I'd have to apologize to everyone at dinner that night and prepare a lesson for his adult Sunday school class that week on why it's important to tell the truth.

I was mortified. I did what he said, but I'll never forget the humiliation and the sense of isolation I felt. It was like I'd been branded with a scarlet letter L for liar, and I felt as though no one treated me the same for weeks after that. That may be the last time I ever really stepped out of line.

The well-intentioned attempt to show Maria the consequences of skipping school and violating the rules of the school and her family registered as:

The support I have from my family and community is fragile. It can be withdrawn in a moment if I don't please them. I have to earn their approval.

This was not the intended or the appropriate message, yet it's what Maria carried through her life, measuring her success by the approval of others. Before she could consider standing up to Jay's pressure, she had to unlearn the lesson—now more than three decades old—that did not serve her and change her repertoire of responses to other people's disapproval.

The most sensitive of approval junkies are reluctant to take any action that might be in their own best interests if they'd risk incurring *anyone's* disdain. Eve, for example, could hardly stand a frown from a store clerk and, as most of us have done on occasion, would agree to keep items she'd gone to return if the person behind the counter made her feel guilty about it. She couldn't risk even a stranger's disapproval.

THE ANGER AVOIDER/PEACEMAKER

Many of us live as though there were an 11th Commandment—"Thou shalt not get angry"—and a 12th—"Thou shalt not get other people angry with you." At the first sign of disagreement, many of us jump to make peace, putting out the fire we fear will blaze out of control.

The peacemakers' desire to provide a measure of calm and rationality in difficult situations can be problematic when it becomes a rigid belief that nothing is worse than a fight. That makes them afraid to argue, even with a friend, fearing that the relationship will break irreparably in two. After all, they tell themselves, surrender is a temporary concession to achieve a greater good.

The Voice of Reason

Liz, who is grappling with the punishing blackmail of her husband, Michael, has the soothing voice of a late-night FM DJ and a manner so placid that it's hard for someone who doesn't know her to see that she's upset. When I remarked on that, she laughed.

Oh, that's my camouflage. When I was a kid I realized watching my brothers and sisters that the ones who yelled back at my mom when she was raging got hit or punished, and the ones who didn't got ignored. I think I must've figured out that you can calm people like you can calm an animal, just by stroking them and talking to them in a gentle way, never getting upset. On my job evaluations they always used to say things like "unflappable" and "great under pressure," and I feel like I have this real talent for defusing tension—like the bomb squad. I like that about myself. For one thing, it makes me totally unafraid of anger, because I know I can handle it and keep it from getting out of control.

When Liz described herself this way, she was pretty convincing because "calming," "soothing," "unflappable" and "great under pressure" had become part of her self-definition. They're qualities she seemed to radiate. But obviously her situation with Michael was far from calm.

I think I fell in love with Michael because he and I are so different. He's outgoing and energetic and direct—he has a real passionate side, a lot of fire. And I'm softer, more in the background. I guess I always knew that he might have a temper, but I never saw it for the longest time, and like I said, I know how to handle anger. That sounds ridiculous, doesn't it? Here I'm married to this raging maniac who threatens me, and I'm scared out of my mind of him and I'm saying I can handle anger. Well, I thought I could. And then it all spun out of control, and so did I. Everything I did, all the stroking and soothing and apologizing, seemed to make him madder. I don't get it. What went wrong?

Liz had spent most of her life refining a style of relating to people that seemed to work for her, one that is highly valued—our volatile society respects people who can rein in tempers. Her soothing voice, manner and approach had been so successful at defusing anger that she erroneously saw herself as a person who didn't fear anger at all—because she knew how to neutralize it. She believed for a long time that if she could just keep the peace, Michael would be nice and she'd be able to reason things out with him. There's no reason to get upset, she told herself. And even when he demonstrated that he was a bully, she resorted to reason, her time-tested tool.

When her well-honed techniques didn't work on him, she felt defenseless and increasingly frustrated. Confronting his escalating pressure and threats had activated a hot button she didn't think was still there, which had its genesis in a childhood full of anger and conflict. As a child she had resolved: "Don't make an angry person angrier. Calm them down or they'll hurt you, or worse, leave you. Don't be the one who upsets them." This decision dramatically limited the options available to Liz, and she'd never learned to express her own anger in an appropriate way. When her calming techniques failed, and she wound up unleashing her own stores of rage and frustration, a crisis quickly mounted.

Until she revisited her fear of anger and broadened the ways in which she allowed herself to respond to it, she would always be vulnerable to blackmailers like Michael and startling eruptions of her own pressurized emotions.

The Other Face of Anger

My client Helen, the literature professor we met in Chapter 1, thought that she'd found the perfect man in her boyfriend Jim. She *knows* she's sensitive to anger, and that's shaped her thinking about the kinds of people, especially romantic partners, she'd like to be around.

I wouldn't ever consider seeing a man who raised his voice to me. My mom and dad gave me my fill of that when I was growing up. Pop's a rebel, which made him exactly the wrong kind of guy to choose a career in the military. He couldn't get promoted to save his life, so he wasn't much more than a file clerk for 20 years. He couldn't stand seeing stupid jerks—his words—being promoted over him just because they were good at conforming, and he was frustrated. He'd come home and yell at Mom, and she'd yell back, and they'd slam doors, bang pots and pans around in the kitchen—it was pretty scary for us kids. I knew nothing was going to happen, but my brother would run into his room and start to cry, and we'd push his bed against the door for protection—so the screamers couldn't come in, I guess. When it got really bad, Dad would storm off and stay away for a couple of days. It was no big trauma, but you know, I don't need that kind of drama in my life anymore. Been there. Done that. It really turns me off.

Helen's adult strategy for avoiding anger—"I prefer not to be around people who get angry"—echoed her childhood strategy—run away and hide till it blows over, or hide in a place where it will never find you. What hadn't figured in Helen's plan was that anger is a normal human emotion, and no matter how diligently she tries to find a place where it doesn't exist, or a person who doesn't express it, she's bound to fail.

When I met Jim, I thought I was in heaven. He's quiet and gentle, and he's always writing me notes and making up songs for me—a real romantic. From the moment I met him, I couldn't imagine him yelling or ever making a scene. Sold! I'll take him! But you know that saying, "Be careful what you wish for because you might get it?" Well, now I know what it means.

You'd think that the way to get to me would be to yell—that makes sense. But Jim is the opposite, and when he gets mad, he gets more quiet. *He won't tell me what's wrong—he won't say anything. I almost wish he'd start yelling so I'd know what I was up against. This is the worst. When he withdraws, I just die inside. It's like I'm totally cut off—like I'm on an ice floe in the middle of the Arctic Ocean. I can't stand it when he gets mad in that quiet, icy way. I have to get him out of that shell, even if I have to stand on my head.*

Or, as had been happening more and more, give in to emotional blackmail.

I helped Helen reconsider choices she'd made, most of them as a child, about how to cope with anger—and then we worked on making a place for it in her life. She was able to make major improvements in her life with Jim, a process we'll see in the chapters that follow.

None of us likes anger, but if we believe that it's always up to us to avoid it, or to squelch it to keep peace at any price, the range of actions available to us is about as wide as a tightrope: we can back down, give in, placate—all the things that tell blackmailers how to get what they want from us.

THE BLAME-TAKER

I encourage people to take responsibility for what they do. But many of us think that we have to take the blame for every problem that comes along in our lives or others', even though we had little or nothing to do with creating it. Blackmailers, of course, feed this notion. In fact, they demand that we buy into it. If they're displeased, we're the problem. And our compliance with them is the solution.

Bizarre Blaming

Eve's world fell apart when Elliot took an overdose of prescription drugs after one of their fights. Elliot was kept under observation in a nursing facility for several weeks, and when he returned home, he blamed Eve for his pain, his problems and all of his fears.

He went off the deep end and he wouldn't stop telling me that everything was all my fault. He said, "See, now they're going to put me in a mental hospital and I'm going to have to kill myself, and you can thank yourself for that. Now I've got this record, and they're going to put me away, and I'm going to be dead because of it." It was horrible. I felt like I was causing him to suffer just for being who I am, and I didn't know what to do.

By any objective standards, Elliot's behavior was ludicrous and his accusations far-fetched. It seems almost inconceivable that a bright young woman like Eve would take him seriously. Yet she *did* believe him. She was sure that what he predicted would probably happen—and all the blame would be hers.

When I asked Eve if she had any idea why she was buying into all this blame peddling, she immediately went to her relationship with her father and we struck oil the first time out.

"My father was always talking about dying," she said. "I think he was obsessed with it." Then she described an incident that took place when she was eight years old.

I'll never forget that day. I can see it like it was yesterday. I was in the front seat of our huge old Pontiac with my father, driving down the street. We were stopped at a crosswalk, and I was looking out the window at some little kids playing in their yard. Daddy turned to me and said, "You don't know anything important, do you?"

I looked at him. "If I had a heart attack right now, you wouldn't know what to do, would you?" he said. "You wouldn't know what to do, and I would die right in front of you." Then he started the car up again and we drove on. He didn't say anything, and I didn't say anything. I just counted the dots on the skirt of my dress and tried not to think about anything at all.

But of course, little Eve did think about what she perceived as her father's indictment of her: You're eight years old, you should be able to save me, and you can't. Eve believed that she was responsible for keeping her father alive—*shouldn't* she be able to do that?—and would therefore be to blame if he died. For a child, the family is the world, and to let them down is to let the world fall into pieces, carrying everyone with it.

"The most true thing in the world in my family was 'If you don't be nice to Daddy, he'll die,'" Eve said. "I really believed it." Eve's father's behavior was bizarre, and to a child it was terrifying. How could she objectively evaluate Elliot's behavior when the bizarre was the norm for her?

Her experience with her father planted the seeds of a style of blame-taking that has carried through strongly to the present. Though you can't always make such direct connections between a person's childhood and their adult difficulties with blame-taking and emotional blackmail, in Eve's case the similarities were blatant.

The Atlas Syndrome

People with the Atlas syndrome believe that they alone must solve every problem, putting their own needs last. Like Atlas, who carried the world on his shoulders, they weigh themselves down with the burden of fixing everyone else's feelings and actions, hoping to atone for past or future transgressions.

Karen, the nurse we met earlier, developed the Atlas syndrome as a teenager, when her father and mother divorced.

When Dad left, Mom was almost completely alone, and I was supposed to make up for the loss. Her family was all back in New York, and we were in California, and she only had one or two close friends out here and she depended on me.

I remember when I was 15 or so, I had a chance to go out on my first New Year's Eve date with a friend. My mother and I had planned to go out to dinner that night, and then a movie, but around Christmas a friend of mine called and said she had a blind date for me, and we could double-date. I was so excited, I really wanted to go, but I felt a little guilty. So I talked over the situation with an aunt of mine who told me, "Frances can't possibly expect you to stay with her when you have a chance like this—go!"

I got up all my courage and told my mother that I wanted

to go out on this date. And she was so hurt, her eyes filled with tears and she said, "Well, what am I *going to do on New Year's Eve?" I did go out, and I had a good time, but when I got home, my mother was in bed with a migraine, practically screaming in pain, and I knew it never would've happened if I hadn't gone. I felt so guilty I could hardly stand it. I didn't want to give up my whole life, but I didn't want to hurt her any more than I had already.*

Karen was only 15, but she had learned to let her mother depend on her. If she didn't take care of her mother, after all, who would? It never occurred to her that her mom could take care of herself. Besides, if she angered her, or "harmed" her by not doing what she wanted, her mother might leave, too.

At first, I didn't know what to do for her, but one day I knew what would help. I got out a pencil and paper, and I wrote out a contract: "I hereby promise my mother that when I grow up, I will see that her life is wonderful. I will see that she has lots of interesting friends and fun things to do. Love, Karen." I gave it to her one afternoon, and she smiled at me and said I was a good girl.

Many of us take on the task of maintaining another person's well-being, an overwhelming responsibility that nonetheless carries its own rewards. Karen had found a way to feel powerful. She had discovered how to make her mother happy—and make sure that her own world didn't fall apart.

It's hard to ignore someone's Atlas tendencies. Karen's daughter, who was blackmailing her by reminding her of the pain she'd suffered in a car accident years ago, had seen the way her mother responded to her grandmother—and most of the other people in her life. It was a snap for Melanie to activate the responsibility hot button.

Melanie and I are close, and I know how hard it is for her to work the Program and hold on to her sobriety. If not for the accident, the scars, she would've been stronger. I'm a nurse, and I know the face of pain. I wish I could've spared her from it, and since I couldn't, I have to protect her now. It's my duty as a mother. I don't like the pressure she puts on me, but I want her to have what I didn't. I love her and my grandchildren so much! Do you know she threatens to not let me see them when she's upset with me? Our family needs to be together, and if I'm the one who has to pull us together, I will.

Like many people with the Atlas syndrome, Karen had no idea where her responsibility for other people began and ended, because she'd long been told that she was supposed to take care of everyone but herself.

Blame and responsibility are closely allied, and it's hard to see the boundaries between them. As I worked with Karen to disconnect her automatic "You're right—I'm to blame. I owe you reparations," she began to define, for the first time in her adult life, how to make room in her world for her own needs, and just what degree of responsibility she wanted to take for those around her.

THE BLEEDING HEART

Compassion and empathy inspire kindness, even noble deeds, and we have little respect for people who lack them. It's hard to imagine how *these* traits could be troublesome. But compassion can turn to a sense of pity so overwhelming that it moves us to renounce our own well-being for the sake of another person. How many times have we said "I can't leave him because I feel so sorry for him" or "She looks at me with the tears running down her face and I'll do whatever she wants" or "I'm always giving in to her, but she's had such a terrible life . . . "? We get caught up in the other person's emotional needs and lose the ability to assess problems and see how we can do the most good.

What gives some people the ability to feel empathy for problems or suffering and offer appropriate help, while others, the bleeding hearts, feel compelled to fly in like Superman and give their all to stop the suffering—even if they have to sacrifice their self-respect or health? As we've seen so far, where there's a compulsion to act and an automatic response that may rebound against us, there's often a hot button at the root.

The Power of Pity

My client Patty, the government worker we met in Chapter 2, grew up in a house that was none too happy for most of her childhood years. Her mother suffered from what seemed to be a deep depression and often disappeared into her bedroom for hours or even days at a time. Patty often jokes, "My mother slept through my whole childhood," but she remembers always being conscious of her mother's presence and her needs, trying to play quietly so as not to disturb her.

I've always been independent, but I was worried about her. Other people's moms weren't sick all the time, but the slightest disruption would send my mother to bed. I really tuned in to her—I could tell by the sound behind the door of her room if she was awake or sleeping, and I could tell if she was sleeping well or if she was restless. If she was sound asleep, I'd look in on her, and I listened to her breathing to be sure she was OK. It was part of my job when my dad wasn't around.

It was a tailor-made training ground for a bleeding heart. When we're in close quarters with a parent or another important person who is physically or emotionally needy, we become ultrasensitive to their cues. Every flutter of the eye, small sigh or change of voice becomes freighted with meaning, and we even tune in, as Patty did, to the nuances of a sleeping person's breathing. A child like Patty can't do anything to help, however.

As we've seen, many of us resolve as children to make things better when we're adults, and it's common for us to reenact scenes from our childhood to make sure they come out right, now that we have the power to fix them.

You know that cliché "You marry your father"? I married my mother! Joe obviously isn't as depressed as my mother was—in fact, what I love about him is how energetic he is when he's happy. But he's so moody—up and down. He has the same way of sighing, even going off to lie down in his room when he's upset, that my mom did, and when he does, all that old training of mine kicks in. Joe talks about how he thinks I can read his mind. When he starts to look so sad, I seem to be able to figure out what the problem is or bring it to the surface the way no one else can. When we first got together, I liked the feeling that we clicked and I could make him happy. But he started to expect me to be a mind reader and it's getting old.

Being with him is like being with a little kid in a toy store. You know how some kids pick up something expensive—that you didn't come for—and cling to it like it belongs to them? And then they look like you've taken away their best friend when you put it back on the shelf? I'm the one who would just buy the damn thing to make the kid smile. Is that so awful?

There are big payoffs for being a bleeding heart, the one who can bring happiness to a poor suffering soul. You escort the other person from the depths of despair back to the land of the living, an almost mythological journey. The joy of "helping" often blinds us to the fact that so much pity-evoking behavior is manipulative: Give the sufferers what they want and voilà! they're cured.

Again, ironies abound when sufferers meet targets who are bleeding hearts. The targets feel helpless in the face of suffering, so rush in to stop it. But by saying yes to every tear-stained demand, they become even more helpless, unable to stop the personal suffering that comes of ignoring their own needs.

The Good Girl Syndrome

When Zoe looked back at her life for a place where a hot button might have formed, she didn't find any particular traumas. Her childhood was happy, she said, and her family was supportive.

The only reason I didn't fit in perfectly was that I wasn't as quiet as girls were supposed to be. I was really competitive, and I have always liked to win. That made my parents pretty upset. When I did well in school, they called me a show-off.

My sisters got the message, and they never seemed to push themselves very hard, but I was just different. My family always said they were proud of me, but it wasn't ladylike to call attention to myself in such a loud way.

Zoe spent years keeping her head down, trying not to "come on too strong" in environments that were slow to encourage women to contribute. But her work didn't go unnoticed, and though she'd never expected to become a manager, she now has 10 people working for her.

The path has been so rough for women, and for me personally, that I've always sworn that I will do things differently. I think there's plenty of room for decency and compassion in the working world, and I've always wanted the people under me to think of me as a friend as well as a boss. I'm not interested in throwing my weight around or imposing my will on my team. We're colleagues, not master and slave. Who says you have to leave your humanity at the door when you've got a corner office?

Zoe had always taken pride in her ability to advise and support other women. In *that* arena, she was comfortable. Zoe the noble and compassionate. Zoe the mentor. Zoe the ever-available friend. She unapologetically excelled at being a bleeding heart, and she didn't want to leave what she considered to be her best qualities behind as she moved up the ladder.

In her determination to be a good person as well as a boss, Zoe made friends with some of her staff, especially Tess. The two women had dinner together regularly and often went to the theater, a passion they shared. Because of their social relationship, it was especially difficult for Zoe to "play boss" with Tess and to say no to her. As we saw with Charles and Sherry, being involved with someone on both a professional and personal level, even if it's a friendship and not a romance, is always tricky and usually ends badly, especially when one person has more authority than the other.

In Charles and Sherry's situation, the boss was the emotional blackmailer, an expected and typical scenario. But in

Zoe's case, the boss had some very sensitive hot buttons, which set her up to become the target of her employee's blackmail.

She just won't let up on wanting me to give her more responsibility. She says I'm supposed to be her friend, and how can I not help her? When I try to reason with her that friendship has nothing to do with my responsibilities to the company, she tells me that I've let my position go to my head and I'm just on a power trip. Oh boy, does that sound familiar to me. I don't want people to be afraid of me or think I'm unfeeling. AAGH! It's making me crazy!

Zoe had not yet resolved the conflict inside herself between the part of her that wanted to be successful and the part that cared more about being liked. She was suffering from the Good Girl syndrome—a malady that affects many contemporary women who still harbor deep concerns about their ability to be powerful and successful and still be loved. Because of her ambivalence about how she was "supposed" to behave, Zoe opened the door wide to blackmail and Tess walked right in.

Tess found Zoe to be an ideal dumping ground, willing to listen to her endless litany of complaints. But when Zoe had something pressing to do, or wasn't able to spend time with Tess, she would be reminded: "You're the only one who can help me. I'm not going to make it without you." This was music to Zoe's ears. This was how she had gotten love in the past—with her caring, compassion and warmth, always being there for people who needed her. But the music was full of sour notes for anyone who wanted to avoid emotional blackmail. Zoe needed to broaden her definition of compassion to include herself.

THE SELF-DOUBTER

Knowing that we're not perfect and that we're capable of making mistakes is healthy. But healthy self-evaluation can easily become self-deprecation. In the face of criticism from someone else, we may disagree at first, then come to believe that our sensors and gauges are faulty. How can we be right if someone important to us says we're wrong? Maybe we're just deluded. We know what we see and experience, but we don't trust it, and frequently we discount the truth of our own ideas, feelings and insights, letting others define how we should be.

This is common when we interact with authority figures, especially our parents. It's "Father Knows Best" all over again. But it can also happen with a lover or a friend we admire, who happens to be a blackmailer. We invest these idealized beings with power and wisdom and believe that they're smarter, wiser, more "right" than we are. We may not like what they do, or see the fairness in what they ask of us, but lacking confidence in ourselves, we let them have their way, never questioning their demands or their version of reality. (This is particularly true for women who got the message early on that they were creatures of emotion and therefore couldn't know anything important, while men were superior, the masters of reason and logic.)

If we assign wisdom and intelligence to another person, which we're bound to do if we don't trust ourselves, it's simple for them to keep our self-doubt active. *They* know best, and what's more, they know what's best for *us*.

When Knowledge Feels Dangerous

Self-doubt can take the form of the statement "I know what I know, but I can't know it." Our knowledge feels uncomfortable, dangerous, and we feel we can't face the changes we'd have to make if we accepted our perceptions as true.

For my client Roberta, who'd been severely beaten by her father and then pressured by her family when she decided to tell, it was painfully difficult to hold on to her reality. "The whole family says I'm wrong," Roberta told me. "What if they're right? How can I be the only one who's right? What if I've just imagined all this? What if I've just exaggerated it?"

Abuse victims often rely on self-doubt to insulate themselves from the horror of their pasts. Among the statements I've heard most often are "Maybe it wasn't as bad as I thought," "Maybe I'm overreacting," "Maybe it didn't happen at all," "Maybe it was just a dream." Roberta needed to cling to reality, but at times her grip was tenuous.

I can't lose my whole family because of this. My whole life I've tried to do something important for them to notice me, but they never did. My brother was the apple of my parents' eye because he was the first son, but when I came, I was just a chubby little girl, and my father couldn't handle that. He hated me from day one. Everything I do is wrong. No one believes me. I just want them to like me, and they hate me now. I must be crazy to go through this. Maybe they're right.

Under the pressure from her family to recant or face exile, Roberta nearly buckled. She had become the family scapegoat.

It's not uncommon for one person to become the representative of everything that goes wrong in a family. Roberta was the repository for the family's denial and secrecy, and she had to absorb its blame, tension, guilt and anxiety in order to keep everyone else in balance. That way, no one in the family had to look at how unhealthy they were.

It's especially tough to believe that your own perceptions are valid when people you love are telling you how crazy, wrong or sick you are, but with support and hard work, Roberta *did* find the courage to hold her ground; her recovery would have been impossible if she hadn't shed the self-doubt that had clung to her for so long. Like all of the behavioral styles we've looked at, it hadn't kept her safe—it had locked her in prison.

Your fight to hold on to what you know, or even just to realize that you've locked away the perceptive center of yourself, may not be as dramatic as Roberta's, but it's every bit as important. Just as owning her own truth was a matter of psychic survival for Roberta, for most of us it's the only way to end emotional blackmail.

A MATTER OF BALANCE

All of the behavioral styles we've seen are survival mechanisms that we've chosen to keep ourselves safe. The problem is, most of them are antiquated, and we've never stopped to review and update them. When kept in balance and alternated with other behavior, none of these styles dooms you to the status of "preferred target" of an emotional blackmailer. Avoiding conflict, making peace, even feeling a little self-doubt won't hurt you—as long as you don't make them the armor that's supposed to protect you from feelings you think you can't stand. If you're a peacemaker but don't compromise when you feel strongly that you cannot do what another person asks—no problem. But if you consistently let these traits run the show, you're holding on to a tow rope that will pull you straight into a sea of emotional blackmail.

TRAINING THE BLACKMAILER OURSELVES

Emotional blackmail takes training and practice. Who provides the training? You do. Who else could tell a blackmailer with absolute certainty and precision: This is what works on me. This is the kind of pressure I always give in to. This is the tool that was custom-designed to probe my most sensitive spot.

You probably don't remember giving a private tutorial in emotional blackmail, but emotional blackmailers take their cues from our responses to their testing, and they learn from both what we do and what we don't do. The following lists will help you identify how accomplished you've become as a personal trainer for the blackmailers in your life.

When faced with a blackmailer's pressure, do you:

- Apologize
- "Reason"
- Argue
- Cry
- Plead
- Change or cancel important plans or appointments
- Give in and hope it's the last time
- Surrender

Do you find it difficult or impossible to:

- Stand up for yourself?
- Confront what's going on?

- Set limits?
- Let blackmailers know that their behavior is unacceptable?

If you answered yes to any of these questions, you are acting as coach and co-star in the blackmail drama. Every day of our lives, we teach people how to treat us by showing them what we will and won't accept, what we refuse to confront, what we'll let slide. We may believe that we can make another person's troublesome behavior disappear if we just ignore it or don't make a fuss, but the message we send when we're not forthright about what's unacceptable is "It worked. Do it again."

It Starts with the Little Things

What many of us don't realize is that emotional blackmail is built on a series of tests. If it works on a small scale, we'll see it again in a more significant arena. When we give in to pressure or discomfort, we're providing positive reinforcement, a reward for bad behavior. The hard truth is that every time we let someone undercut our dignity and integrity, we are colluding—helping them hurt us.

We live with the illusion that emotional blackmail springs full-blown into our lives like a whirlwind, coming from nowhere and knocking us over with its force and fury. "How could the other person have changed so dramatically?" we ask ourselves. "How could things have suddenly become so strained?" And sometimes the entrance of emotional blackmail into a relationship *is* sudden, but just as often it builds slowly, gaining ground over time because we let it.

Liz began telling me about her problems with her husband, Michael, by describing how frightened his threats of punishment made her feel. But as she looked back, she realized that long before the big crisis between them, she had been letting Michael get away with numerous incidents of more subtle emotional blackmail.

Michael's always been Mr. Perfect. He's one of those guys who'd make a date to meet you and then leave if you were even five minutes late—just to make sure you knew you should've been on time. I should've had a clue when he started lining up my magazines on the coffee table and complaining when they weren't just so. His rules—and he does *have a rule for every little thing—were a source of tension between us from the moment we started living together. And when we had the twins, forget it. Just try to keep a house spotless with toddlers. But reality meant nothing to Michael. He kept letting me know he wanted the house kept a certain way. How did he let me know? He had his ways.*

I remember one day when I left a few dishes in the sink instead of putting them in the dishwasher. When I came home, Michael had piled them on the floor. I could hardly believe it—but I didn't say anything. I just swallowed hard and picked them up.

Liz jumped to the conclusion that she'd been wrong and deserved Michael's anger, but she was the one doing the training. Michael couldn't fail to notice how effective his punishment had been.

Now that I think about it, it seems like he was always finding a way to correct my lapses. Once I left without closing the garage door, and when I got home, Michael had unhooked the automatic door opener so I had to get out of the car and open the door myself. It was like one of those punishments your parents dream up that you're supposed to never forget. He had me convinced that I was a slob, irresponsible and a bad mother, and I felt guilty as hell and ended up apologizing.

Petty punishments like Michael's strip us of our adult dignity and power. They're the emotional equivalent of spankings, and they work to infantilize us—reduce us to bad little children who need to learn a lesson. As Liz found, the guilt we feel easily becomes a sense that "I'm bad, so I deserve this treatment."

With her hot buttons buzzing, Liz didn't even consider letting Michael see how upset she was, and she never thought of confronting him. But by hiding her feelings, she was training him to escalate his punishing behavior to keep her in line. Blackmailers learn how far they can go by observing how far we *let* them go. We don't know what would have happened if Liz had nipped this behavior in the bud. We do know Liz taught him that if he infantilized her, insulted her and threatened her, he'd get what he wanted. Michael's punishments were repeated, ever escalating, finally reaching a painful and frightening apex when he threatened to cut her off financially and take the children if she tried to leave him.

On the surface there doesn't seem to be much of a relationship between disconnecting a garage opener and the bolder threats that came later, but the early incidents were like the cold that can lead to pneumonia—dangerous when ignored and untreated.

As all targets find, in emotional blackmail the present is a prologue to the future. What you teach today will come back to haunt you tomorrow.

SELF-BLACKMAIL

Despite the title of this chapter, sometimes blackmail only takes one. We can easily stage every element of the blackmail drama alone—from request through resistance, pressure and threats, acting as both blackmailer and target. This happens when our fear of other people's negative responses is strong, and our imagination takes over. We assume that if we ask for what we want, they'll disapprove, withdraw, be angry, and we're so adamant about protecting ourselves that we don't let ourselves take any chances, even the minimal risk of asking the other person "How would you feel if I . . ."

Let me show you what I mean.

My friend Leslie has been dreaming about taking a trip to Italy for more than a year, making arrangements with friends and lining up tickets for the opera. But six months ago, her daughter, Elaine, went through a rough divorce, and Leslie has been helping her out with an occasional loan and by babysitting for Elaine's two small children. Leslie and Elaine have been through their share of rough times, but they've gotten closer since Elaine's divorce, and Leslie is thrilled about the new friendship that's bloomed between them.

"I couldn't possibly do anything to jeopardize that," Leslie told me, "and I just know that if I take the trip, she'll get angry and think I'm being selfish. How can I go on this vacation when she's struggling and needs my help?" Leslie's daughter would probably cope perfectly well if her mother told her about the trip, but Leslie refuses to give herself a reality check, preferring to postpone a much-needed holiday.

How often do we deprive ourselves of something that's reasonable and well within our means simply because we fear another person's reaction? We shelve our dreams and plans because we're "sure" someone will object—though we've never tried bringing up our ideas. We want something, we resist, we pressure ourselves by making up the negative consequences, and we keep ourselves from doing what we want to do. We create our own FOG. That's self-blackmail.

We may have a history with other people that justifies our apprehension about their reactions, but often we make completely unrealistic assumptions. We may even come to resent people for keeping us from doing something they didn't know about. We step carefully around our hot buttons and lock ourselves in a safe, airless pattern of self-blackmail.

A NOTE OF CAUTION

Please don't use this chapter as a way of beating up on yourself. Until now, you've done the best you could with what you knew. You've been a member of the PTA—that vast group of people whose behavior took shape Prior To Awareness. Look compassionately at the person you've been, and then use this chapter to give yourself a deeper understanding of the emotional blackmail transaction and the role you have played in it.

7

The Impact of Blackmail

Emotional blackmail may not be life-threatening, but it robs us of one of our most precious possessions—our integrity. Integrity is that place inside where our values and our moral compass reside, clarifying what's right and wrong for us. Though we tend to equate integrity with honesty, it's actually much more. The word itself means "wholeness," and we experience it as the firm knowledge that "This is who I am. This is what I believe. This is what I am willing to do—and this is where I draw the line."

Most of us would have no problem listing the do's and don'ts or "Thou shalt" and "Thou shalt not"s that guide us. But weaving those beliefs into the fabric of our lives, and defending them under the pressure of emotional blackmail, is far more difficult. Many times we capitulate and compromise our integrity, losing our ability to remember how it feels to be whole.

What does integrity feel like? Take a moment to look at the following list—you might want to read it aloud—and imagine that each of the statements is true for you most of the time.

- I take a stand for what I believe in.
- I don't let fear run my life.
- I confront people who have injured me.
- I define who I am rather then being defined by other people.
- I keep the promises I make to myself.
- I protect my physical and emotional health.
- I don't betray other people.
- I tell the truth.

These are powerful, liberating statements to apply to ourselves, and when they genuinely reflect our way of being in the world, they give us a balance point, a sure sense of equilibrium that keeps us from being pushed off center by the stresses and pressures that are constantly coming at us. When we give in to emotional blackmail, we cross off the items on this list, one by one, by forgetting what's right for us. And each time we do, we sacrifice a little bit more of our wholeness.

When we violate this essential sense of ourselves, we lose one of the clearest guiding forces in our lives. We're set adrift.

THE IMPACT ON OUR SELF-RESPECT

Wimp. Gutless wonder. Failure. Chump. We have dozens of ways of describing ourselves as we walk away from an emotional blackmailer, having given in once more. Our judgment about ourselves becomes cloudy as we endure the blackmail FOG: "If I had any backbone, I wouldn't be giving in" we say to ourselves. "Am I really this weak? What's the matter with me?"

There's no need to be inflexible or to beat yourself up if you give in to someone over relatively minor issues. Most of us realize that we often have to bend a little and make compromises, and there are plenty of times when giving in to pressure doesn't mean that much. But falling into patterns of giving in to things that aren't good for you takes its toll on your self-image. There is always a bottom line, a point at which to give in is to violate our most important principles and beliefs.

How We Let Ourselves Down

The price we pay for acting as though that point does not exist became painfully clear to Maria over time. Some months after we began working together, she was unusually quiet in one of our sessions, a change from her typically outgoing manner. When I asked her what was going on, she answered slowly.

I'm angry about many, many things. Of course I'm not happy about what Jay's done. But what bothers me the most is what I have done to myself. I know that we've talked a lot about family, family, family and how I've always honored that and made it come first in my life. But when I look in the mirror, I see a woman who has not respected herself enough to tell her husband, "I will not let you degrade me or my marriage by being unfaithful." I feel as though I've really let myself down. I have done everything but stand up for myself. I might as well be wearing a sign that says, "Kick me."

I pointed out to Maria that, though it didn't seem like it to her, she'd come a long way, and I reminded her that she'd worked hard to reach a point where she recognized her own needs and could push back against the pressures of her upbringing and environment. Part of the intense self-reproach she was feeling was the result of seeing clearly, for the first time in years—or maybe ever—that she'd managed to have a strong, conscious sense of values that respected and protected the rights of everyone but herself.

A Vicious Cycle

Honoring and protecting our integrity isn't easy. Blackmailers shout down our inner guidance by creating confusion and uproar, and as they do, we seem to lose contact with the knowing parts of ourselves, only to kick ourselves when we realize we've given in again.

Patty, who let her husband, Joe, persuade her to ask her hospitalized aunt for a loan, epitomized the plight of the target who's being heavily pressured to give in to blackmail.

It was a totally no-win situation. If I didn't make the call, I would have felt like just this horrible, horrible person letting Joe down. He's the breadwinner, and here he is asking me this little favor when he does everything. It seemed so reasonable. But after I'd done it, I felt just awful . . . horrible, worthless. I felt used, like I had no spine, which was true.

Patty was caught in the classic "damned if you do, damned if you don't" dilemma that drowns so many blackmail targets in waves of self-condemnation. As long as she accepted Joe's premise that what he wanted so urgently was a "little favor" that she owed him, she couldn't bring herself to resist doing what he asked—even though, as she put it, "I'm not the kind of person who would do something like that. Who in their right mind would call someone who just had surgery and beg for money?"

Patty didn't lose her sense of what was appropriate, but in her need to keep the peace with Joe, she acted as though she had. As a result, she was filled with remorse and self-contempt.

The unfortunate result of such self-flagellation, however, is that it creates a vicious cycle. Under pressure, we do something that doesn't fit with our sense of who we are. In shock and disbelief, we realize what we've done and begin to believe that we are actually as deficient as blackmailers make us out to be. Then, having lost our self-respect, we're even *more* vulnerable to emotional blackmail because now we're especially desperate for the approval of our blackmailers—which would prove that we're really not so bad. We may not be able to uphold our own standards, but maybe we can meet theirs.

As Patty put it,

I was afraid that if I didn't make the call, he wouldn't love me, that I wasn't a good wife. And I needed him. He wouldn't love me, and I would be letting him down.

Even though Patty felt terrible about making that phone call, it was more comfortable for her than saying no to Joe. Given a choice between violating her own sense of right and wrong and looking like a bad wife, we know what choice she made.

Rationalizing and Justifying

Protecting our integrity can be frightening and lonely. It puts us at risk of incurring the disapproval of people we care about, and it may even jeopardize a relationship. Margaret wanted to hold on to Cal—no matter what—so she did what many blackmail targets do given a choice between being true to themselves and complying with what someone else wants: she rationalized.

Margaret tried to come up with "good reasons" for doing what Cal wanted. She told herself that group sex was no big deal—maybe she *was* being old-fashioned and prudish. After all, he was so wonderful in so many other ways. Margaret's need for such extensive rationalizing should have been a tip-off to her that she was stepping outside the boundaries of what she knew to be true and healthy.

It takes a lot of mental and emotional energy to persuade ourselves that we can accept something that's not OK for us. An internal war rages between our integrity and the blackmailer's pressure, and as in any war there are losses and casualties. Margaret paid an agonizing price for her rationalizations, and we worked together to rebuild her self-respect, silence the criticism she was blasting herself with and strengthen her connection to that guiding voice inside.

No matter how confused, self-doubting or ambivalent we are about what's happening in our interactions with other people, we can never entirely silence that inner voice that always tells us the truth. We may not like the sound of the truth, and we often let it murmur just outside our consciousness, not stopping long enough to listen. But when we pay attention to it, it leads us toward wisdom, health and clarity. That voice is the guardian of our integrity.

Eve had signed up for classes that would help her get a job and gain some financial security as she worked to build her art career. But under pressure from Elliot, her optimistic plan crumbled.

All I want to do is get some skills so that I don't always have to be dependent on someone. I thought I'd take some computer graphics courses, a little bit of illustration, so I don't always have to hope for some big commission to come along.

But he really, really hated that, and one day, literally the day I was going to take this silly computer test, he threatened to overdose. I was so stunned, it was almost like my ultimate nightmare had come true. He was sitting there with a bottle of liquor and a whole lot of bottles of pills. How could I go back to school after that? I told myself, Don't do it, Eve, stay in the classes, but . . . I just . . . collapsed. I just said, Look, to hell with it.

Like most blackmail targets, Eve lost sight of the fact that some of the most important promises we make are to ourselves. Compared with Elliot's pressure, and her belief that she was responsible for his very survival, those promises seemed almost trivial.

Elliot's threats were dire, and Eve wasn't ready yet to deal with them, but even when the pressure is less intense, many targets will cop out on themselves. One of the most serious effects of emotional blackmail is the way it narrows our world. We often give up people and activities we love in order to please our blackmailers, especially if they are controlling or overly needy.

But every time you don't take the class you want to take, every time you stop pursuing an interest or stop seeing people you care about in order to make a blackmailer happy, you are giving up an important part of yourself and diminishing your wholeness.

THE IMPACT ON OUR WELL-BEING

Emotional blackmail leaves us full of unexpressed smouldering feelings. Patty was highly resentful of Joe's manipulations—and her feelings were perfectly natural—but even though she was aware of how she felt, she couldn't get the relief that would come from venting her anger and frustration. Most blackmail targets tend to stuff these feelings, only to have them surface in all kinds of distressing forms: depression, anxiety, overeating, headaches—an entire spectrum of physical and emotional manifestations that take the place of directly expressing the way we feel.

When Catherine's therapist used emotional blackmail to pressure her into joining a group, Catherine was full of anger, not only at the therapist but also at a close friend who had been drawn into the situation.

My friend was in the group and she started to pressure me, too. Later I found out that Rhonda had told her to work on me. So I had both of them. I really felt double-teamed and very angry. But it was anger I didn't feel safe to express directly. In fact, I didn't even know if I had a right to be angry. So I just got more depressed instead. The whole experience was really awful.

Rhonda did me a lot of harm. I was very vulnerable at the time, and she never supported or validated me. She made me feel worse about myself—more inadequate and more unlovable. Thank God I had the wisdom to see what was going on and get away.

Like Catherine, many targets of emotional blackmail question their right to even *have* certain feelings—especially angry ones. They may turn their anger inward, where it can become depression, or they may rationalize to cover up how angry they are. Catherine was fortunate because she finally was able to move past her depression and self-doubt and get out of an unhealthy situation.

When Mental Health Is on the Line

Eve was so enmeshed in her destructive relationship with Elliot that she believed her sanity was being threatened.

I knew I had really gotten myself in trouble. My emotions were so frazzled I was afraid they were gonna have to take me in and put me in the rubber room. I literally had to get in a rocking chair and rock. It really felt like I was just gonna go bananas, and I couldn't get any emotional distance from him. It was that horrible combination of rage and love and guilt.

When emotional blackmail is as oppressive and omnipresent as it was for Eve, it creates emotions of such intensity that we do sometimes believe we are "going crazy." I assured Eve that many people confuse strong emotions with going crazy and that there was much we could do to diminish those fears for her. She was right—there did need to be some emotional distance before she could begin to deal effectively and calmly with the soap opera her life had become. And together we would create it.

As Eve so clearly illustrates, emotional blackmail can be dangerous to your mental health. It can also be dangerous to your physical health, especially if you try to push yourself past your physical limits to please a blackmailer.

Physical Pain as a Warning

Kim, the magazine editor who was working punishing hours in response to pressure from her boss, woke up in the middle of the night with shooting pains from her shoulders to her wrists.

I feared something like this was coming, but it's always a shock when it hits you. I don't know why I couldn't say, "My arms are beginning to hurt and I have to slow down a little—do one person's work instead of two or three people's." But I hear Ken's voice in my head telling me how great Miranda was, and I'm determined to prove to him that I'm just as good. And that son-of-a-bitch knows just how to get to me. The really scary thing is, I did this to myself.

When we don't protect our bodies, they signal us with pain until we pay attention. For Kim, repetitive strain injuries mean that there may be crippling consequences for giving in to pressures to work too hard.

In Kim's situation, cause and effect were pretty obvious. Too much work, long hours and the pressure to excel exhausted her, and her body rebelled.

I certainly don't go along with the idea that every physical ailment is psychosomatic, but there is ample evidence that the mind, emotions and body are intimately connected. Emotional distress can significantly increase our vulnerability to headaches, muscle spasms, gastrointestinal problems, respiratory disorders and a host of other ailments. I believe strongly that the stress and tension that accompany emotional blackmail may manifest themselves in physical symptoms when other outlets of expression are blocked or shut down.

BETRAYING OTHERS TO PLACATE THE BLACKMAILER

We know that capitulation to emotional blackmail causes us to betray ourselves and our integrity. But we tend to overlook the way that, by trying to calm blackmailers or avoid disapproval, we may also be betraying people we care about.

We've seen many examples in this book of how blackmail affects the other people in the target's life. Josh is betraying Beth by telling his parents he's not seeing her anymore, and she is deeply hurt. She feels unprotected by him. She also knows that when the time comes, as it always does, and the truth comes out, there will probably be even more uproar than there would have been if he had dealt with the situation in a more courageous way in the first place.

Karen found herself caught in the middle between her mother and her daughter, faced with having to hurt one or the other.

I had been planning a little party to celebrate Mom's 75th birthday. When Mom asked who was coming, I ran down the list, but when I got to Melanie she stopped me in my tracks. "I don't want Melanie there," she told me. "I know she's your daughter, but she's been just terrible to me lately—absolutely disrespectful. The last time I called her, she was too busy to even talk to me. She's just nice when she wants something."

I tried to smooth things over and tell her that Melanie has had a lot of things on her mind lately, but Mom wouldn't hear it. "If you don't tell Melanie not to come," she said, "I don't want your party. You can have it without me. I've spent other birthdays by myself, and I can spend this one alone too." So I had to tell my own daughter that she wasn't welcome at her grandmother's party.

Karen allowed herself to get pulled into the conflict between her mother and Melanie, and she became the conduit and messenger for all the bad feelings between two adult women. Like most of us, she hadn't learned any effective strategies for dealing with emotional blackmail, and she assumed that she had only two choices—to give in to her mother and hurt her daughter or to stand fast and risk hurting her mother—a real lose-lose position.

Many of us have been in an impossible situation like Karen's, asked to choose between two people we care about, because of a particular blackmailer's needs. "It's your kids or me" is a common demand. That was the choice Alex gave Julie when he decided her son was taking too much attention away from him.

Another familiar scenario may involve various family members who pressure one another to ally with one parent against the other, especially after a divorce. If the divorce has been a bitter one, a typical form of emotional blackmail will be "If you continue to speak to your father/mother you're out of my life [or my will] and I'll never speak to you again." It's a painful dilemma. Whatever choice the target makes, someone will be betrayed, adding to the already heavy load of guilt and self-reproach.

THE IMPACT ON THE RELATIONSHIP

Emotional blackmail sucks the safety out of any relationship. By safety I mean goodwill and trust—the elements that allow us to open ourselves to someone without fear that our innermost thoughts and feelings will be treated with anything but care. Remove those elements and what's left is a superficial relationship empty of the emotional candor that enables us to be our true selves with another person.

As the safety level of a relationship drops, we become guarded, hiding more and more from our blackmailers. We stop trusting them to care about how we feel or what's best for us or even to tell us the truth, knowing that when they're out to get their way they can be at best insensitive and at worst ruthless. The victim is intimacy.

Shutting Down

Eve talked about the erosion of intimacy in her relationship with Elliot in an especially poignant way:

I know he sounds very bizarre and crazy, but it wasn't always like this. The first year we were together it was a totally different relationship—simple and romantic. He's bright and incredibly talented and we really loved each other. It was only when I moved in with him that he showed his crazy side.

Now it's like a pressure chamber. I can't even describe it. It's the way you feel for someone you'd been angry at and maybe they get some awful disease, but you love them and deep down care about them. But there's no intimacy, genuine intimacy. I'm not talking about sexually. I'm talking about emotionally. I can't tell him my real feelings, because he's so vulnerable . . . fragile. I can't tell him my dreams or plans, because he's so threatened by them. They're not safe subjects. There's no real intimacy when you have to watch everything you say.

Targets of emotional blackmail become so accustomed to negative judgments, disapproval, pressure and overreactions that, like Eve, they are reluctant to share major parts of their lives. We stop talking about:

- Dumb or embarrassing things we did. Our blackmailers might ridicule us.
- Sad, frightened or insecure feelings. Our blackmailers might use them against us to prove we're wrong to resist what they want.
- Hopes, dreams, plans, goals, fantasies. The blackmailer might shoot them down or use them as proof of how selfish we are.
- Unhappy life experiences or difficult childhoods. The blackmailer might use them as evidence of our instability or inadequacy.
- Anything that will show we're changing and evolving. Blackmailers don't like it when we rock the boat.

What's left when we must consistently walk on eggs with someone? Superficial small talk, strained silences, lots of tension. Just below the artificial calm that surrounds a placated blackmailer and a target who's given in is the widening chasm that's opening between them.

Karen's mother resorts to arm-twisting to get Karen to spend more time with her, yet for all the closeness that's left between them, she'd do just as well to talk to a cardboard cutout of her daughter. The rigid interaction between them has no room for the real Karen or what matters to her. It's as though there are coils of barbed wire separating these two women, one strand consisting of the mother's criticism and the other of Karen's efforts to protect herself by withdrawing.

It's amazing how much of ourselves we hold back when we're trying to stave off another blackmail episode. We keep crossing the street conversationally to avoid a serious topic, or worse, a demand. Zoe put it well:

I don't even ask Tess how she's doing anymore because I know she'll tell me, and then she'll want me to make it all better for her. I figure we can talk about, oh, the weather, the Dodgers, Mel Gibson, movies—but only comedies. I just keep it light.

In a blackmail-tainted situation, relationships with friends, lovers and family members that once had real depth begin to get thinner as the roster of safe topics shrinks.

Allen, the businessman, believed that he had to be careful about what he shared with Jo because of her extreme dependency and overreactiveness.

I can't tell Jo I'm feeling scared or insecure, because I'm supposed to be the rock of Gibraltar here. But she's my wife, and I wish I could just share even a little of what I've been going through lately. I'm having trouble with my business—sales are way off—and I've had to dig into some investments to cover the bills. There's a small factory in San Jose I want to go take a look at. They're talking about some new contracts that could be a lifesaver. But I don't even want to mention making a trip and being gone for a few days. She'll freak out. And I can't tell her the truth, because she panics. Hell, this isn't a real partnership—it's a one-man band.

Allen censored himself from talking about the topics he thought Jo "couldn't handle," and as a result, even though they lived together, he felt very alone, without the intimacy that comes from being able to share the dark times as well as the bright ones. Their marriage was in an emotional straitjacket.

Constricting Emotional Generosity

In one of the great paradoxes of emotional blackmail, the more we feel the blackmailer demanding our time, our attention or our affection, the less we feel free to give. We frequently hold ourselves back from expressing even casual affection because we're afraid it may be misinterpreted as a sign we've given in to their pressure. We turn ourselves into emotional misers, not wanting to feed the blackmailer's hopes or fantasies.

Roger, the screenwriter, talked about this paradox early in my work with him, before he and Alice had gotten their relationship onto more solid ground.

Alice and I have a lot of terrific times together, and I'd really like to be able to tell her that and let her know how much I appreciate her and what a wonderful woman she is in many ways. But I can't say anything that sounds too loving, because I know she's going to hear it as a marriage proposal. Or she'll start on the baby thing again. I'm a very affectionate person, but I find myself holding back a lot—I don't want to mislead her—and then I feel lousy because I don't feel free to express myself, and I know that she feels rejected.

Roger, at that time in their relationship, didn't feel free to express his real feelings—even though they were positive—because he knew that whatever he said would wind through the loops and convolutions of Alice's unrealistic expectations and turn into ammunition for future emotional blackmail.

Often we have to guard our happy feelings as well as our loving ones because there's no reason for celebration unless you can arrange to be happy on the blackmailer's terms. Josh obviously can't share any of his joy with either parent, because his father's powerful disapproval of Beth has made it unsafe to do so.

"He doesn't want to hear about it. I'm not supposed to lead my own life. He says he loves me, but how can he?" Josh asked. "He doesn't even know who I am."

The relationship Josh's father thinks he has with his son doesn't exist. Obedient Josh doesn't exist. And what is real—the contentment that Josh has found with Beth—is only allowed to exist outside his father's knowledge. The relationship between father and son is a sham, and so are many, if not most, of our long-term dealings with emotional blackmailers.

When safety and intimacy are gone from a relationship, we get used to acting. We pretend that we're happy when we're not and say that everything is fine when it isn't. We act as though we're not really excited about something when we are, and we pantomime loving the people who are pressuring us, even though we hardly know them anymore. What used to be a graceful dance of caring and closeness becomes a masked ball in which the people involved are hiding more and more of their true selves.

It is now time to take all this understanding and turn it into action so that you can start to deal effectively with emotional blackmail and the people who are using it against you. You will be amazed at how quickly you can reclaim your integrity—and dramatically improve your relationship with the blackmailer.

PART II

Turning Understanding into Action

INTRODUCTION: A TIME FOR CHANGE

One of my favorite stories is about a man who's driving down a road when he sees a woman crawling around on her hands and knees under a street lamp. Thinking she must be in trouble, he pulls over.

"What's going on?" he asks. "You look like you could use some help."

"Thanks," she replies. "I'm looking for my keys."

After they search for awhile he asks, "Do you have any idea where you dropped them?"

"Oh, yes," she replies. "I lost them about a mile down the road."

Understandably baffled, he asks, "Then why are we looking here?"

And she answers, "Because it's familiar territory and the light's better."

A lot of us think we can solve our problems with emotional blackmail by sorting through a repertoire of familiar behavior to find an escape. We accept the blackmailers' accusations, buy their blame, apologize and ultimately comply. There's a certain logic to this—we know how to do it, and compliance brings immediate relief. But if we stick to our accustomed ways of responding, we'll never lay hands on the real keys to ending emotional blackmail. They're a mile down the road, in the self-affirming, nondefensive behaviors I will teach you in this part of the book.

It's vital to move from the comfortable, well-lighted arena of habitual responses into the far more uncomfortable sphere of behavioral change. By now you have a good understanding of how and why you're being blackmailed, but knowledge is utterly useless if it doesn't become a catalyst that moves you to do what it takes to end the blackmail. Change requires using information, not just collecting it. To change, we have to know what we need to do and then *we have to act.* Yet for many reasons, most of us resist this step with every fiber of our being. We're afraid we'll try and fail. We're afraid we'll lose the good parts of our relationships when we try to get rid of the bad. We, who are good at so many things in our lives, often resist changing self-defeating behavior patterns with exquisitely wrought reasons why we can't do anything different.

So we wait to learn new behavior until we feel a little less anxious, a bit less afraid or insecure. And the blackmail gets worse. The good news is that if you're willing to take action now and let your feelings of confidence and competence catch up with you, you can end emotional blackmail. The bad news is that you have to begin the change process while you are still afraid.

STEP BY STEP

If you want to deal effectively with a blackmailer, you have to learn some very different responses and communication skills. The words that come out of your mouth have to change, be replaced by a new style of response and expression. The emotional tone surrounding your responses has to be different. You have to interrupt the ritualistic pattern of resistance, pressure and capitulation by changing the reactions that have kept you on automatic pilot.

In this portion of the book, I'll walk you through a process I've designed to take you step by step from where you are now to the point where you can respond in a new way the next time someone tries to blackmail you. I'll teach you powerful nondefensive communications skills and guide you through visualizations, checklists and writing exercises that can quickly bring change on many levels.

We'll be proceeding along two tracks. The first, which you can start to use almost immediately, is the behavioral track. At first you may feel as though you haven't changed at all inside—it's quite likely that you'll still feel guilty or obligated or afraid when a blackmailer turns up the pressure. But you will learn to act more effectively, and once you change your behavior, the relationship must change as well. The results you'll see will embolden and encourage you.

At the same time, we'll work together on the emotional track, which will take you through the somewhat longer process of changing your inner world, disconnecting old hot buttons as well as doing some work on the wounds and erroneous belief systems that have made you vulnerable to emotional blackmail.

It may sound odd for someone who's been a therapist for 25 years to tell you this, but you can do a lot of this work successfully on your own. Obviously, if there is abuse in the relationship or if you are suffering from deep depression, unmanageable anxiety, extreme degrees of self-hatred or a dramatic lack of confidence, there are wonderful resources for you to take advantage of, and this book can serve as an important adjunct to one-on-one or group therapy, 12-Step programs or personal growth seminars. In most cases, though, all you'll need is courage and determination.

In the past, you have behaved in automatic, predictable ways when faced with emotional blackmail. You've argued, tried to explain your position, offered up some active or passive resistance and ultimately given in. Now it's time to replace that way of behaving with a far more self-respectful, efficient and empowering set of techniques. With practice—the willingness to keep using these techniques until they become comfortable for you—you will put an end to blackmail.

As you're exploring the chapters that follow, you're likely to face blackmail situations that will give you an immediate opportunity to apply what you're learning as you learn it. Jump in and take your new skills off the page and into your life. I promise you that when you see yourself begin to deal in a more conscious way with blackmailers' pressure, you'll notice a dramatic improvement in the way you feel about yourself.

Once you are less afraid and feel less manipulated by fear, obligation and guilt, you'll notice how many choices open up to you. You'll be able to decide who you want to be close to, how much you are responsible for other people, how you really want to use your time and love and energy.

Please be patient with yourself, and be persistent. Some of you may start this work feeling that your self-respect and integrity have been so profoundly violated that they are lost to you forever, but I urge you to use the word *misplaced* instead of *lost*—and then look to new behavior to help you find them. Together, we'll do the work of rediscovering and rebuilding what emotional blackmail has worn away inside you and in your relationships with blackmailers. I applaud you for taking concrete steps to eliminate emotional blackmail from your life.

8

Before You Get Started

You've probably heard the joke about the tourist in New York who stops a man with a violin under his arm and asks for directions to Carnegie Hall. "You want to know how to get Carnegie Hall?" replies the violinist. "Practice, practice, practice."

We all know those familiar directions, and most of us are able to recognize the relationship between practice and mastery in many aspects of life. You probably remember wobbling toward steadiness on a bicycle or awkwardly moving your fingers on a keyboard while learning to type.

But when it comes to making important changes in our lives, we often expect results overnight. The unavoidable truth is that learning new skills takes practice, and it may be awhile before you're comfortable using them. Just as we have to walk around in a new pair of shoes before they really fit, we have to break in new behavior. You probably won't see immediate changes in your life the first day you make the commitment to free yourself from emotional blackmail—but you will see them soon. Remember, commitment is a promise to yourself, and it's one well worth keeping.

THE FIRST STEP

Before you think about dealing with the blackmailer, there are some things you need to do yourself. Every day for the next week, I'd like you to set aside some private time to work with three very simple tools: a contract, a power statement and a set of self-affirming phrases. You'll need as little as 15 minutes a day. I'd like you to take the phone off the hook, remove yourself from interruptions and focus on *you*. Some people have found that the only private time they have is in the bathtub, in the car or at their desk during lunch. That's fine. You can do this work anywhere.

The first thing I'd like you to do is sign a contract that lists a number of promises I'd like you to make to yourself—ground rules for this process. You may have serious doubts at this point about your ability to keep promises like these, especially if you've tried unsuccessfully in the past to stop giving in to emotional blackmail. I'd like you to put the past aside right now and begin to take a new set of steps based on new understanding and new skills.

This contract is a powerful symbol that puts your willingness to change into tangible form and helps clarify your goals.

Some people find that they get the best results from handwriting this contract on a sheet of paper. You may also want to write it on the first page of a notebook you devote specifically to the exercises I'll be teaching you. If you want to record your observations and feelings as we go, please do.

Whether you recopy the contract or simply sign it in the book, please read it aloud to yourself every day this week.

Second, I'd like you to learn and practice saying a power statement, one short sentence that you can use to keep yourself grounded when blackmailers turn up the pressure.

Power statement: I CAN STAND IT.

Those four words may look insignificant, but used correctly, they can become one of your most potent weapons for resisting emotional blackmail. They're effective because they counter a belief that ushers us straight in to saying yes to our blackmailers: the idea that we *can't stand* the pressure.

CONTRACT WITH MYSELF

I, ____________, recognize myself as an adult with options and choices, and I commit myself to the process of actively getting emotional blackmail out of my relationships and out of my life. In order to reach that goal, I make the following promises:

- I promise myself that I am no longer willing to let fear, obligation and guilt control my decisions.
- I promise myself that I will learn the strategies in this book and that I will put them into practice in my life.
- I promise myself that if I regress, fail or fall into old patterns, I will not use slips as an excuse to stop trying. I recognize that failure is not failure if you use it as a way to learn.
- I promise to take good care of myself during this process.
- I promise that I will acknowledge myself for taking positive steps, no matter how small they are.

______________ Signature

________________ Date

- "I can't stand to hurt his feelings."
- "I can't stand it when she says things like that to me."
- "I can't stand my guilt!"
- "I can't stand my anxiety."
- "I can't stand it when she cries."
- "I can't stand his anger."

We are constantly making statements like these to ourselves, and if you truly believe you can't stand it—whether you're referring to a blackmailer's tears, a red-faced shout or a "gentle" reminder of how much you owe someone—you'll only be able to see one course of action. You'll have to back down, give in, go along, keep the peace. This belief is the fundamental trap that undermines blackmail targets. We've made "I can't stand it" our mantra, and in effect, we've brainwashed ourselves. Though you may not believe me now, you are a lot stronger than you think. You *can* stand the pressure, and your first step is to replace any belief that tells you otherwise.

Repeating "I CAN stand it" will begin putting a new message into your conscious and unconscious mind. For this week, every time you think about taking steps to eliminate blackmail and start to get frightened, upset or discouraged, stop and repeat this statement to yourself. Breathe deeply, exhale completely and say, "I can stand it." Do this at least 10 times.

I suggest practicing by imagining yourself face-to-face with a blackmailer who's in the midst of pressuring you. Have you seen the clear acrylic shields that riot police sometimes use? Let "I can stand it" be a shield that comes between you and the blackmailer's words and nonverbal expressions. Say the power statement aloud as you practice. You may be timid with it at first, and not very convincing, but stay with it. You'll start to believe yourself. Is this process mechanical? Yes. Does it feel foreign? Yes, it might. But remember that your old responses have not worked. I assure you that repeating "I can stand it" to yourself *does* work.

REVERSING THE BEHAVIORS OF COMPLIANCE

Now, using the same basic concept of replacing old beliefs with new ones, I'd like to help you develop a set of self-affirming phrases that will calm you, make you feel stronger and embolden you to act. First, let's take a look at a set of statements that describe the typical feelings and behavior of emotional blackmail targets toward their blackmailers. Most, or even all, of the statements may be true for you—not in all your relationships, certainly, but when you face blackmail. Check all that apply.

When dealing with emotional blackmailers:

__I tell myself that giving in is no big deal.
__I tell myself that giving in is worth it if it gets the other person to shut up.
__I tell myself that what I want is wrong.
__I tell myself it's not worth the battle.
__I give in now because I'll take a stand later.
__I tell myself it's better to give in than to hurt their feelings.
__I don't stand up for myself.
__I give away my power.
__I do things to please other people and get confused about what I want.
__I acquiesce.
__I give up people and activities that I care about to please the blackmailer.

These statements sound pretty feeble, don't they? But don't be embarrassed. Until a few years ago, most of the statements would have been true for me, too, in certain relationships—and they're true for many people. Emotional blackmail is very prevalent, and we're all in this together. Pay attention to how the statements you've checked make you feel, and use the following list to help you pinpoint the full range of feelings that go along with this behavior. Circle the words that apply to you, and add any other feelings you're aware of that I haven't listed.

How do I feel when I act this way?

Embarrassed	Frustrated	Martyred
Hurt	Emotionally numb	Agitated
Ashamed	Sad	Scared
Angry	Powerless	Resentful
Weak	Self-pitying	Victimized
Depressed	Helpless	

If you circled angry, I wouldn't be surprised if you were angry at yourself, and even angry at me for reminding you of certain aspects of your behavior that you'd rather forget. Use this discomfort—it's letting you know what aspects of your behavior need attention.

Now take your original list of statements, and change each item you've checked to its opposite. For example:

Old: "I tell myself that what I want is wrong."
New: "I ask for what I want, even when it upsets the blackmailer."

Old: "I give in now because I'll take a stand later."
New: "I hold my ground and take a stand now."

Old: "I do things to please other people and get confused about what I want."
New: "I do things to please myself as well as others, and I am clear about what I want."

You can also put your original behavior statements in the past tense by saying, "I used to [statement], but I don't do that anymore." For example: "I used to tell myself that what I want is wrong, but I don't do that anymore."

Try both approaches and see which feels best to you. Then repeat these new positive statements aloud as if they described you. I know they're not true right now, but they'll give you a sense of what it will feel like to be free of behavior that's driven by fear, obligation and guilt. Putting these statements in the past tense or restating them in a positive way helps drain power from them and return it to you. Some of my clients have found this exercise to be especially effective when they repeat the positive statements to themselves while looking themselves in the eye in a mirror. It will give you a chance to literally see yourself describing your actions in an affirmative way.

Think about how you would feel if you were to act in this new way. Use the following list to help you describe those feelings.

Strong	Proud	Confident	Courageous
Elated	Triumphant	Excited	Hopeful
Self-affirming	Powerful	Capable	

These adjectives will help you visualize yourself dealing confidently with blackmail. Change starts with a vision, and it's important to give yourself a clear mental picture of what you are trying to achieve. Then, as we work, you can energize your vision with action and move steadily toward your goal. You may want to write or repeat a statement that expresses this vision—"I stand up to emotional blackmail and feel strong, confident, proud and elated."

Please go through these lists every day for a week as you mentally review your recent and past interactions with emotional blackmailers, and each time you do, jot down five or six feelings that come up for you. You'll probably notice that your feelings change over time, and you may find that it becomes harder for you to repeat the negative statements and easier to imagine yourself resisting blackmail.

After a week of working by yourself with these three exercises, you should be feeling more centered and ready to start dealing directly with your current situation. Let yourself take the time to do the preparatory exercises, no matter how eager you are to get going. You have plenty of time—neither your blackmailer nor the blackmail is going anywhere.

SENDING UP AN SOS

I'd like to give you a game plan, starting with the first steps to take before responding to a blackmailer's demands. It's easy to remember, and stripped down to the essentials it's *When you feel as though you're sinking under the pressure of emotional blackmail, send up an SOS.*

You don't need know Morse code or wave flags around. Just remember this convenient shorthand for the first three steps in the change process. SOS: Stop. Observe. Strategize. We'll cover the first two steps in this chapter and explore tools and strategies in the following chapter. Don't skip any steps—building your strategies on a strong foundation is essential.

STEP ONE: STOP

Patty was a little bewildered when I told her that the first thing any target of emotional blackmail has to do is *nothing. That means you don't make a decision about how to respond the moment a demand is made.* This sounds easy, but it can be challenging—especially when the pressure to give an answer feels intense—so it's important to brace yourself and prepare.

You'll feel awkward at first. That's OK. Feel awkward—and keep going.

How do you do nothing? Well, the first thing you need to do is give yourself time to think—away from the pressure. To do this, you'll need to learn some time-buying phrases that will slow things down. Below, I've listed some suggestions for your first response to whatever the demand happens to be:

- I don't have an answer for you right now. I need some time to think.
- This is too important to decide quickly. Let me think about it.
- I'm not willing to make a decision right now.
- I'm not sure how I feel about what you're asking. Let's discuss this a little later.

Use time-buying statements as soon as a demand is made, and continue to repeat them if the blackmailer pressures you to make an immediate decision. How much time should you ask for? Obviously, the higher the stakes and the more the profound the issue, the more time you need. You can make a quick decision about where to spend a vacation or whether to buy a computer, and even if your choice isn't great, not that much is lost. But when it comes to such major life issues as the fate of a marriage, children or changing jobs, you should be prepared to take as much time as you need to think things through.

As part of our learning process with this book, when a blackmailer pressures you about something that falls outside the "major issue" category, let him or her know you want at least 24 hours to make a decision. You'll be using the time you've bought for yourself to make a decision and prepare to stand behind it.

Your Timetable or Theirs

What makes emotional blackmail unique is the sense of a clock constantly ticking in the background. There's a request on the table, and at some point you need to respond. A lot of the blackmailer's pressure comes from the idea that there's no time to lose. It's the same illusion that makes thrillers and suspense films so effective—the story is set up as a race against time. We get caught up in that drama and don't bother to question whether it's real. If you do step back, you'll see that in the vast majority of cases there *is* no urgency, except in the mind of the blackmailer.

When you step uncritically into the "act now, last chance" universe, which is where almost all blackmail is set, the pressure is on. Time-buying statements allow you to turn off the sound of the ticking clock and view this drama from the outside. Maybe the sale on cars or computers only lasts until Sunday—but there will be another sale. Maybe the blackmailer does have an important deadline—but it's not your deadline.

You have something the blackmailer wants. Time is on your side. When you use a time-buying statement, all you want is time to think—something almost any reasonable person would be willing to give you. But some blackmailers will even use blackmail to get you to back down from your time-buying. That was one of Patty's concerns.

"This is all fine," she told me as we practiced the time-buying statements. "But you just don't know Joe. As soon as I tell him I want time to think, he'll start pouting, and then he'll say, 'You know the sale ends this week. It's not like we have all the time in the world. What's the problem?'"

"And what will you say?" I asked.

"I'll try something like, 'I'm not willing to make a decision right now,' but I know that's going to go in one ear and out the other. He's going to be like a kid—'How long will it take? How long will it take?'"

"And you'll keep repeating, 'It'll take as long as it takes,'" I told her. The blackmailer may resent you for wanting time, pout or use other forms of pressure, but the power of repetition is generally enough to deliver the message that you're serious.

A New Dance

These time-buying statements may confuse or upset the blackmailer. After all, you're changing the usual and expected scenario by not automatically giving in. It's almost as though the two of you had been doing a tango when all of a sudden *you* began to waltz. The blackmailer may experience your slowdown as resistance or a negative answer, and as a result may start pressuring you immediately. The result can be, and often is, chaos. Just by saying "I need time," you've shifted the balance of power in the relationship and put the blackmailer in the position of waiting to see what you're going to do—a reactive, and for them a much less powerful, role.

Be prepared for even more pressure as blackmailers try to regain their position. As they stick to their by now familiar script, stick to your new set of responses and mentally repeat "I can stand it."

The force of old habits and the blackmailer's skillfulness at creating FOG can make using the new responses an unnerving experience. With punishers, who dislike giving up any measure of control in a relationship—and with any blackmailer who's strongly resistant to your time-buying statements—it's important to clarify your motives. You might want to say something like:

- This is not a power struggle.
- This is not about my trying to control you.
- This is about my needing more time to give thought to what you want.

If you are dealing with a rational person, these are thoughtful, reassuring statements that can help defuse tension.

Doing It Right but Feeling Bad

Because asking for time and stating your motives takes you out of your accustomed role, it's not unusual to do everything right and still feel as though you've bungled your end of the transaction. That's what Zoe found:

You want to know what happened? It was terrible. Tess was pushing me to put her on our big new shoe account because the partners who own the firm will be in from New York in a week and she wanted to impress them, especially Dale, who she's convinced has his eye on getting rid of her position. The visit was a week away, and she wanted to have the assignment NOW. So she tried everything. She'd say, "I know I'm going to get fired if you don't let me do this job, and I don't know what I'll do if you don't help me. I hate to pressure you, but I really need help right now." Her eyes started to tear up a little.

I did what you said. I said, "I'm sorry, but I can't make an important decision like that right now." And she came right back with, "But you know how much this means to me. I really need your help. Aren't we friends? Don't you trust me? You know I'll do a good job, and you know I'd do this for you."

By this time, I'm getting caught in this "Oh my god, oh my god. I have to help her. This is urgent. She's right. If I don't help her now, she'll be in big trouble. I have to do something." I could feel my heart starting to beat faster and I felt like I was breathing faster, too. I tried to slow it down, and I did say a couple of quick I-can-stand-it's. And then I said, "I know you want me to do something right now, but I need time to think about this. We can talk tomorrow."

She gave me an angry glare and said, "I thought you were my friend as well as my boss. I thought friendship meant something to you." And then she walked out and I totally felt like shit. And I still feel like I let her down. I thought this was supposed to make me feel better. I feel awful.

"Congratulations," I told her. "That means you're breaking old patterns." Bad habits are comfortable, seductively so—until you feel their consequences. Buying time may not be easy, but it gets easier as you go along. And as I reminded Zoe, *all* you are doing at this point is delaying a decision. You haven't done anything but put the blackmailer on your timetable—hardly the drastic step the blackmailer will probably say it is.

As you continue to use your stalling scripts, the blackmailer may respond with increasing desperation. The operative words are "Gimme—NOW."

Learning to tolerate discomfort in the service of healthy change is one of the most difficult things any of us has to do. In the past, discomfort has always been the prelude to compliance, but now you're changing all that, and you're going to feel unsteady. It's OK to feel uncertain and anxious as you regain your integrity. Things are starting to shift internally as well as externally, and I want to reassure you that it's perfectly natural to feel shaky when that's happening.

Don't let discomfort throw you off course.

A Dialogue with Discomfort

Zoe felt increasingly uncomfortable about clinging to her request for time. Tess was insistent, and she became the picture of suffering whenever Zoe saw her. The more she thought about Tess's request, the more she realized that she couldn't give in to it—yet the guilt she felt seemed to get ever larger.

I can't get over this feeling that I'm a heartless criminal. I feel worse and worse about this. I'm doing nothing, and it's tearing me apart. Are you sure this is working?

Internal discomfort is one of the major impediments to change, and we're so used to responding to it as though it were a fire to be put out that many of us haven't learned to live with it in the natural amounts that accompany change. We push it away, extinguish it, treat it as though it has no place in our lives—and by doing so, we eliminate some of our most effective options. Most of us are so reluctant to examine our discomfort that we often misinterpret what it's trying to tell us by reacting to its presence blindly instead of asking what it means.

I told Zoe that one of the ways to reach wholeness is to reclaim even the parts of ourselves we're struggling with and learn to recognize them as normal. One means of doing this involves entering into a dialogue with those parts—pulling them outside ourselves and getting to know them. For our next session, I asked Zoe to pick an object in her house that represented discomfort to her—an itchy sweater, an unflattering picture of herself, a tight pair of shoes—and we'd use it to get to know more about the feelings that seemed so formidable to her.

When she came in, I asked her to take the object—she'd picked an unattractive pair of high heels that had never fit her—and put it on the empty chair in front of her. She was then to start talking to this symbol of her discomfort as though she were talking to a person. Later, she'd play the part of the discomfort itself and let it talk to her.

Zoe had never done anything like this before, and she felt understandably self-conscious and somewhat reluctant. But I explained to her that she could learn a lot about this state of being that seemed to have so much power over her. I encourage you to try this yourself, and feel free to vent. Tell the discomfort just how you feel about it and ask it questions.

Zoe warmed up to the process once she got going. Here are some of the things she said.

Discomfort, you think you're really hot stuff, don't you. You've been running the show for a long time and I'm sick of it. I've given you a lot of power, but I'm putting you on notice that those days are over. I thought you were bigger than I am—and maybe you knew more—but when I look at you, I can see that you're small, you're ugly and you get me into trouble. In fact, whenever you take over, I'm such a wimp and a coward I don't even know who I am anymore. I'm really tired of you. Is there any good reason why I shouldn't just show you the door?

I asked Zoe what the experience had felt like for her.

I felt kind of silly at first, but as I got into it, I realized that this part of me really takes over a lot of the time. It's just one part, and I act as though it's a 500-pound gorilla. It's a lot more like these shoes—it just doesn't fit in my life as much as I thought it did.

For the next part of the exercise, Zoe sat opposite me in the chair and held the shoes as she played the part of her discomfort, responding to what Zoe had just said.

You show me the door? That's a laugh. I'm not going anyplace. I like it here and I'm not leaving without a fight. It's pretty cushy here—all I have to do is make one tiny squeak and you jump to let me do whatever I want.

Zoe came away from the exercise with a new way to look at a sensation she'd always thought she had no control over and had branded intolerable. But I told her that this awareness was not going to change things for her overnight. Though she'd begun to find some of the keys to her regular patterns of caretaking and capitulation, her discomfort wasn't going to go away without a fight. Her job would still be to change behavior while she was feeling uncomfortable. Meanwhile, she could continue to get to know more about what brought discomfort up for her when she faced a blackmailer—and keep exploring ways to see discomfort as just one element of a situation, rather than the unbearable totality.

I hope you'll use this exercise yourself. You can speak to an object or, if you'd prefer, write a letter to your discomfort and a letter from your discomfort to you. Some people like to type a dialogue, first speaking to the discomfort, then asking questions of it and letting discomfort answer.

Your words and discoveries may be very different from Zoe's, but I'm sure you'll gain valuable information. The point of the exercise is to externalize the discomfort, look at it and begin to develop ways of handling it that go beyond running the other way. When you face discomfort, you'll find that it's smaller, less menacing and far less threatening than it felt when you tried so hard to avoid it.

Turning a Triangle into a Straight Line

There's another version of the do-nothing technique that's useful when you find yourself in the middle of a conflict between two other people, or when a third party is using emotional blackmail on you to benefit someone else. The action you need to take: Get out of the way.

For Karen, in the midst of her mini-crisis with her mother and her daughter, doing nothing actually proved to be the beginning of some real healing for all three women. Karen said:

Let's say I'm able to stall the way you say when my mom says, "If Melanie comes to the party, I want to call the whole thing off." And let's say I'm able to tell her that I can't make a decision about this right away and I'll call her back. Then what?

"Then you call your mother and tell her your decision is that you're not going to make a decision. Look, Karen, this is between your mother and Melanie. You know what happens to the referee when he tries to get between two fighters—he gets clobbered! You need to get out of the middle. Tell your mother that she's the one who's going to tell Melanie that she doesn't want her at the party and you're not going to do it for her. You've got plenty of time to cancel the party if you need to. Let's see what happens."

As expected, Frances howled and complained and pressured Karen to do her dirty work for her, which was to be expected. But when she saw that Karen wouldn't budge, she did indeed call her granddaughter and told her that she was upset with her. To everyone's amazement, this opened the door to some honest dialogue between Melanie and her grandmother, which resulted in some major air-clearing between them—and the beginning of a more honest relationship. This, in turn, had a positive effect on Karen's relationship with Frances because Frances saw her daughter refuse to yield to pressure. As a result, Frances gained a new respect for Karen and began to realize the old manipulations weren't going to work anymore. All this because Karen did "nothing."

It's essential when you get triangled into a conflict between two people in your life that you gracefully remove yourself from the ring by refusing to carry messages or become an arbitrator. If you don't remove yourself, it's almost a given that the bad feelings between the other people will end up being dumped on you, and nothing will get resolved.

In Maria's case, her in-laws put themselves in the middle between Maria and Jay and were pushing hard on her obligation hot buttons to keep her from leaving their son. Maria had already told Jay that she needed time to make such an important decision, and despite his fervent pleas for an answer, she didn't waver. However, with her in-laws she was having a tough time being resolute.

I know how much I'm hurting them, and they don't deserve it. They're dear, sweet people and they haven't done anything wrong, but I know how much they'll suffer if I divorce Jay. His mom calls me almost daily and reminds me how much it would mean to them if Jay and I could work things out.

I told Maria that she, too, needed to learn how to do nothing. In her case, that involved not having repetitive coercive phone conversations with her mother-in-law or dwelling on this subject with anyone else who had their own agenda. A sample set of responses I gave Maria can help you disengage yourself from the pressure of third parties.

Mother-in-law: "Fred and I can't stand this. We don't know what's happening or what's going to happen. We're so worried about you and Jay and our grandkids. How long are you going to take to decide about a divorce?"

Maria: "Mom, I haven't made a decision."

Mother-in-law: "Then how long is it going to take?"

Maria: "Mom, it'll take as long as it takes. Let's talk about something else."

Just keep answering that you haven't made a decision, repeat that it'll take as long as it takes to decide—and then change the subject. People ask us lots of questions, and often we feel we have to answer definitively right away. Of course, we don't. It's fine to answer with "I don't know." It's equally fine to say "I'll let you know when I've made a decision." And if pressure continues, it's fine to steer the conversation to other topics. Even when the person pressuring you is not the blackmailer, but rather a person you like and respect, it's vital to maintain your own timetable and not feel rushed to make a decision, especially an important one.

Take Some Space

Buying time gives you a chance to experience your own thoughts, your own priorities and your own feelings. Remember that you have a lifeline to hold on to—your time-buying statements. You may feel like a broken record by the time you're through repeating them, but keep going and give them a chance to work.

However, if after using these statements you still feel so anxious and pressured that you're tempted to do something to alleviate your discomfort, *walk away.* I'm not talking about turning on your heel and leaving the other person without explanation. I'm talking about excusing yourself and going to another room where you can be quiet for a few minutes. You can say "I need a drink of water" or "I need to use the bathroom—I'll be right back." Or if you feel *really* anxious, how about "I need a drink of water *and* I need to use the bathroom."

By the way, you can do this at home, in a restaurant, at the office, on an airplane, in fact, just about anywhere. Putting some physical distance, even a room or two, between you and the blackmailer can take a lot of urgency out of the transaction and give you some all-important emotional distance as well.

When I talk about emotional distance, I mean turning down the flame and allowing your feelings to cool. When confronted with emotional blackmail, your feelings may be so intense that you can't think, reason, judge or look at what your choices are. Emotional blackmail is in your face: intense, pressured, demanding—full of frantic energy. This cacophony of feelings often seems overwhelming. You're in a state of pure emotional reactiveness, and you need to move into a more cognitive, detached mode. Taking a few minutes to quiet yourself will do that for you. Calm yourself, repeat "I can stand it," and resolve to buy yourself time.

STEP TWO: BECOME AN OBSERVER

Once you've detached yourself from the blackmail drama, you're in a position to gather the information that will help you decide how to respond to the blackmailer. During the time you have bought to make your decision, you'll need to become an observer of both yourself and the other person.

Use Visualization

To help you do this, I'd like you to use this visualization exercise: Envision a glass elevator on the ground floor of a 50-story observation tower. I'd like you to picture yourself inside the elevator as the car slowly starts to move up. As you look out on the lower floors, it's difficult to see anything because of a swirling ground fog. Occasionally the fog breaks and you can make out the outlines of objects and people, but they're vague and fuzzy, appearing and disappearing. This is the realm of pure emotions, the gut feelings that blackmailers churn up in us.

The elevator car keeps moving up, and as it does, you leave the fog behind and begin to see a wider landscape. By the time you reach the top, you have a panoramic view, and you can see that the fog that you thought covered everything is confined to the valley at the base of the tower. What seemed all encompassing was just a tiny patch, a small part of the picture. The elevator has reached a different plane, a place of reason, perception and objectivity. Step out of the car onto the observation platform. Enjoy the quiet and the clarity. Remember that you always have access to this place.

Traveling up from the gut level to the head is useful when you're under the pressure of blackmail because it's easy for us to get so caught up in the feelings of fear, obligation and guilt that our perceptions are fragmentary or distorted. I'm not asking you to detach from your feelings—I'd just like you to add perception and reason to the mix so that you're not driven solely by feeling. Both the intellect and the emotions contain a great deal of information, and we need to create an exchange between the two. The goal is to be able to think and feel at the same time, rather than just thrashing around in the emotions alone. When blackmail heats up, you need the perspective of the observation tower.

What's Really Happening?

Take time alone to think about the blackmailer's request, stepping into the role of observer. Your feelings will still be there, but turn your attention away from them and let your mind review the situation. Ask yourself: What just happened? It's a good idea to write down the answers to the questions that follow. Taking the information out of your head and putting it on paper will also help you gain emotional distance. If not, you can do this entirely in your head. Either way, answering these questions will provide a lot of clarity for you.

First, step back and look at the demand.

1. What did the other person want?
2. How was the request made? For example, was it posed lovingly, threateningly, impatiently? Use any description that applies to your situation.
3. What did the blackmailer do when you didn't agree immediately? Here, you'll want to take into account facial expressions, tone of voice, body language. Be as specific as you can. What were the blackmailer's eyes doing? Where were their arms and hands? Where did they stand while speaking to you? What gestures did they use? What tone of voice? What was the overall emotional tone? Put into words the picture you've taken in.

   Here are notes that Patty took after a recent scene with Joe:

He was withdrawing, sulking, acting very upset. His whole posture and body language was communicating how sad and disappointed he was. His arms were crossed on his chest and he wouldn't look me in the eye. He was sighing a lot and pulling at the lint on his sweater, and when he talked he whined. Then he got up, slammed the door and turned up the radio in the bedroom.

Next, look at your own reactions to the demand.

1. WHAT ARE YOU THINKING?

Write down what's going through your mind, paying special attention to thoughts that repeat or keep intruding. They'll give you valuable insights into the beliefs that you have formed over the years. Among the common beliefs I see in blackmail targets are:

- It's OK to give a lot more than I get.
- If I love someone, I'm responsible for their happiness.
- Good, loving people are supposed to make the other person happy.
- If I do what I really want to do, the other person will see me as selfish.

- Getting rejected is the worst thing that could happen to me.
- If no one else will fix the problem, it's up to me.
- I never win with this person.
- The other person is smarter or stronger than I am.
- It won't kill me to do this, because they really need me.
- Their needs and feelings are more important than mine.

Which of these statements seem true to you? Which do you identify with the most? Ask yourself: Where did I learn this, and how long have I believed it?

None of these beliefs is true, yet we cling to them because they're what we've absorbed over the years. Often, as I've mentioned, we think we chose such beliefs, but they've been imparted by powerful people at every stage of our lives—parents, teachers, mentors, close friends. It's important to identify our beliefs about ourselves as they come up in the face of emotional blackmail because beliefs are the precursor to feelings.

Feelings aren't the ephemeral, independent forces we often think they are. They're a response to what we think. Almost every anxious, sad, fearful or guilty feeling we have in response to emotional blackmail is preceded by a negative or erroneous belief about our own adequacy, lovability and responsibility to others. And these beliefs are fountainheads of our feelings. Our behavior, then, is often an attempt to quiet the uncomfortable feelings that spring from these beliefs. The bottom line: to change self-defeating behavior patterns, we must start with the root element—our beliefs.

When Eve guiltily drops out of school because Elliot is upset, she's acting on a belief that his feelings have more value than hers do. First comes the belief: the other person is more important than I am, and what I want isn't important. From the belief flow the feelings: guilt, obligation, pity. And finally, the behavior: dropping out of school.

We know now that our moods are influenced as much by our brain chemistry as by the events in our lives, but even many of the people who have repetitive depressions and anxiety disorders as a result of biochemical imbalances can exacerbate these conditions with self-denigrating beliefs. Uncovering your deep beliefs can help you learn why you feel the way you do. And once you've done that, you'll begin to see how these beliefs and feelings become the catalyst for self-defeating behavior patterns of compliance and capitulation.

2. HOW ARE YOU FEELING?

What do you feel as you replay your transaction with the blackmailer? Write down as many feelings as you are aware of, using the following list as a starting point:

Angry	Threatened	Hurt	Guilty
Irritated	Insecure	Frustrated	Disappointed
Wrong	Inadequate	Doomed	Scared
Anxious	Unlovable	Resentful	Stuck
Trapped	Overwhelmed		

This check-in is the equivalent of taking your emotional pulse, and though it's simple, it's an important diagnostic tool. Keep in mind that a feeling is an emotional state that can be expressed in one, or at most two, words. The moment you say "I feel *like* . . . " or "I feel *that* . . . ," you're describing what you think or believe. Because we're trying to differentiate between thoughts and feelings, and find the relationships between them, it's important to be clear.

Example: "I feel *that* my husband always wins" is a thought. To get to the feeling, you can say, "I *believe* that my husband always wins, and I *feel* discouraged."

Now check in with your body.

As you look at your list, identify where you feel these feelings physically. Are they churning in your stomach? Knotting in your neck? Gripping in your back? Burning in your cheeks? Notice how your body reacts to your feelings.

Sometimes our bodies will tell us a truth that our minds don't. We may say we're really not anxious—then notice that we're drenched with perspiration. No, no, nothing's wrong—so why is my stomach in knots? The body's responses cut through denial and rationalization, and the body won't lie to you. Keep in mind that any time you identify feelings of anger or resentment, you're being alerted to watch out for aspects of the demand that are not in your best interests.

3. WHAT ARE YOUR FLASHPOINTS?

The blackmailer's words and nonverbal language resonate within us in very particular ways, and it's important to know what our personal triggers are. Facial expressions, tone of voice, gestures, postures, words and even smells can activate belief systems and feelings that lead us toward giving in to a demand. They're the wires that lead straight to our hot buttons, and the more we know what gets to them, the closer we can come to disconnecting them.

Observe yourself, and think over past instances of blackmail. Then list the behaviors that get to you the most. Among the triggers I've seen are:

- Yelling
- Door-slamming
- Particular words (like "big shot," "selfish") that make us feel bad about ourselves
- Crying
- Sighing
- An angry face—red face, eyebrows drawing together, angry scowl
- The silent treatment

Then connect the behavior to your feelings: When the blackmailer does __________, I feel ____________.

When I asked Josh to begin to connect his father's appearance and behavior to specific responses in himself, he noticed that it was his father's appearance even more than his words that turned up the anxiety levels he felt.

"Susan, I made my lists, and I noticed that whenever my father's face gets red—before he even opens his mouth—I get scared. I looked through all the adjectives, and I tried to find something more dignified than *scared,* but that's the one that fit the best. Scared for me is fight or flight—I start going on pure animal instinct."

It's important to be as honest as you can while you're observing. Don't judge the feelings, don't evaluate them, and don't try to decide whether they're valid or not valid or whether you have a right to be feeling them. Turn off the running commentary and observe. I've found that it helps many people to preface their observations with phrases such as:

- Isn't it interesting that . . .
- I'm beginning to notice that . . .
- I never realized that . . .
- I'm becoming aware that . . .

Josh noticed that his defensiveness and anxiety lessened when he phrased his observation this way: "Isn't it interesting that whenever I see my father's face get red, I feel scared." It's a far more thoughtful and objective statement than "I get really scared when my father's face gets red." This objectivity helps put you in a more cognitive mode and helps insulate you from self-criticism. "When I said to myself, 'Isn't it interesting,' I felt a lot less like a baby and a wimp," Josh said.

The "Isn't it interesting" let Josh know that what followed were observer's comments, and it gave him more distance from the judge inside that tends to critique and label so many of our reactions.

Keep observing until you begin to make connections between your beliefs, feelings and behavior. Our blackmailers have made these connections both instinctively and intellectually, and they've used them to keep their advantage over us. But now you've begun to level the playing field, and what was once "insider information" is available to you. Now I'll give you the tools to turn this preparation and knowledge into effective behavioral strategies that will dramatically alter the existing patterns between you and the blackmailer.

9

A Time for Decision

In the past, you've generally responded to the urgency of the other person's needs and desires with automatic compliance—an almost reflexive response to pressure. But now that you've bought some time, you have the luxury of being able to consider what *you* want as well. Although I can't make your decision for you, I *can* help you ask some penetrating questions that will enable you to take an objective look at what's being asked of you and thoughtfully decide whether you want to comply or resist. Once you've done that, I'll show you powerful and effective ways to present your decision to the blackmailer and deal with his or her reaction to the choice you've made.

THE THREE CATEGORIES OF DEMANDS

To begin, I'd like you to go back to the other person's demand and answer some questions about it. Write the answers down, and as you do, don't censor yourself, and don't feel that you're permanently bound to what you put on the page. If you change your mind or have new insights, go back and add, erase or enlarge upon your original response.

- Is something in this demand making me uncomfortable? What is it?
- What part of the demand is OK for me, and what part is not?
- Is what the other person wants going to hurt me?
- Is what the other person wants going to hurt anyone else?
- Does the other person's request take into consideration my wants and feelings?
- Is something in the demand or the way it was presented to me making me feel afraid, obligated or guilty? What is it?
- What's in it for me?

You'll notice that if you look at the component parts of a demand, you may very well be able to accept all but one or two of them. For example, your husband is pressuring you to travel across the country to visit his folks, which you're happy to do, but this is your busy time of year at work, and that's making you uncomfortable. This is important information to have as you shape your response.

You may begin to feel alarmed if you answer yes to the red-flag questions that ask if you or anyone else will be hurt by complying with the demand. That's your integrity barometer warning you of an impending storm.

As you look over your answers, you'll find that most demands fall into one of three categories:

1. The demand is no big deal.
2. The demand involves important issues, and your integrity is on the line.
3. The demand involves a major life issue, and/or giving in would be harmful to you or others.

Different decisions and responses, naturally, are appropriate for each category, and in the sections that follow, I'll help you assess your answers and consider options suited to each one.

NO BIG DEAL

Small decisions come up daily in most of our relationships. We go back and forth about the timing and cost of purchases, where to spend our vacations, how much time to spend with someone, how to balance career, family and friends. These are not matters of life and death, and it's common not to have strong feelings about disagreements in one of these arenas. No matter what happens, no one will suffer great harm, and the primary cause of friction is probably the blackmailer's pressure tactics rather than the content of the demand. Some of us tend to cede this category automatically to the blackmailer, thinking there's no harm in doing so. It's small potatoes. No big deal.

But please proceed with caution here. The word *automatic* is one I'd like you to try to erase in any dealings with an emotional blackmailer. No matter how small the issue, examine the demand, especially the style in which it was presented to you. Get a clear understanding of the parts of it that bother you, if any, and look at the transaction in the context of the relationship as a whole.

Going Through the Process

When Leigh, the stockbroker whose mother, Ellen, constantly used negative comparisons, mentioned that she was having a tough time at work and was dreading the prospect of going out to dinner with her mother later in the week, which Ellen was pressuring her to do, I asked her to go through this process.

"Oh come on, Susan," she said. "This is ridiculous. So I'm tired. It's just one dinner, and it's not going to kill me."

"Just walk through the list," I told her. "You never know what you'll find out."

"OK," she said reluctantly. "This will be fast. The only thing about Mom's wanting to go out that bothered me was that when I said I was tired, she said something about how Caroline always had time for her. I don't mind taking Mom out, and it's crazy to ask if it's going to hurt anyone—of course it won't. Does she care how I feel—well, not entirely, but this is just dinner, for god's sake. Why should I fight about that? Is she making me feel afraid? No. Obligated? Kind of. Guilty? A little. But so what? I'll probably go and be glad I did—believe it or not, we like spending time together. And what's in it for me . . . I'll make her happy and I'll feel good about that."

I asked Leigh how answering the questions had made her feel.

"I guess my neck and jaw are a little tight," she said. She knew from the observation work she'd done that those were anger's tension spots for her, a cue to pay attention.

In contrast to the overreactiveness of many blackmailers, which we saw in Chapter 5, many targets of emotional blackmail tend, like Leigh, to *under*react. That means they often minimize their uncomfortable feelings, denying that anything is bothering them and using rationalizations to persuade themselves that their objections to other people's requests are groundless.

I suggested to Leigh that in the process of looking at what her mother wanted from her she could ask some additional questions to help make her more aware of her habitual ways of responding. I certainly don't suggest that you put every transaction under a microscope—there's no need to overanalyze everything and remove all spontaneity from your dealings with other people. But if you've experienced discomfort and emotional bullying in a relationship, it's important to use a more critical eye than you would in other circumstances. If you think that you may be an underreactor, I suggest that you ask yourself:

- Is a pattern developing here?
- Do I seem to be in the habit of saying "It's no big deal," "No problem," "I don't have a preference" or "I don't care"?
- If it were entirely up to me, what would I do?
- Is my body telling me something different from what my mind is telling me? (For example, you're thinking: It's only a movie, so even though I don't feel like it, I'll go—but you notice that your stomach is pumping more acid than usual.)

If you answer yes to these questions, it's time to take a stand and express your own desires. You may decide to say yes, but you need to identify the elements of the demand that are troublesome to you and make a decision to tell the other person about them. Give yourself permission to say "I don't want to" or "I don't feel like it" without feeling that you have to give elaborate explanations. Don't question your right to say no to something that seems relatively unimportant. Standing up for yourself on the small issues will give you an opportunity to develop the skills you need to hold fast if or when the stakes are higher.

Keep in mind that sometimes what's most objectionable about a demand is the way it is presented. Sometimes style *is* substance, and we can't ignore it. Leigh said:

I don't care—really!—about taking Mom out. What pisses me off is the way she got me to say yes. I hate being compared to Caroline, and I'd like her to stop it.

The pressure blackmailers use can feel insulting, grating or demeaning, and it's vital not to minimize it or let it pass just because the issue on the table seems insignificant and you're not planning to object to it. In Leigh's case, it was most important to call her mother's attention to how resentful the negative comparisons made her feel. Yes, she could take Ellen to dinner because the action itself wasn't a problem, but she needed to explain to her mother how to ask for time with her without using emotional blackmail.

Conscious Compliance

Conscious compliance is the yes you choose after thinking about what another person wants and after you have disabled the mechanisms of automatic compliance by observing and becoming aware of your thoughts, feelings and preferences. Used appropriately, conscious compliance can be the best way to achieve the results that are most important to you. But remember that this form of compliance is the result of a careful reflective process. It follows from the Stop, Observe, Explore steps I've described.

Conscious compliance is a good choice when:

- You examine the demand and find that it has no negative impacts. Perhaps it was delivered with whining or mild sulking, but none of the behaviors that accompany it are habitual, and you and the other person are not frozen in a blackmail pattern. The other person may want you to do something you regard as boring or tedious, but it's not harmful to anyone. You can look at your decision to say yes as part of the natural give and take of a good relationship, an expression of generosity that is likely to be returned.

- You examine the demand and find that it will have no negative impact as long as it involves an even trade with the blackmailer. You will comply this time, but the blackmailer agrees to let you make an equivalent decision next time. For example, I pick this year's vacation spot, you pick next year's. I'm not suggesting that you keep score or reduce your exchanges with a friend, colleague or loved one to "I gave you your way twice and you only did what I wanted once, so you owe me." But if you look back over your recent dealings with another person and realize that you are doing most of the compromising, you are witnessing the beginnings of a power imbalance. It's important to address it before it becomes entrenched.

- You examine the demand and find that you can say yes willingly, and without harm to yourself or others, but only to parts of it. When this is the case, conscious compliance involves striking a deal—saying yes to what you can. In exchange, you ask the blackmailer to drop the elements that you find disturbing.

- You examine the demand and decide to say yes for a time—and you label your compliance as strategy. You know why you are saying yes, and you develop a plan for changing the parts of the situation that are not acceptable to you.

The first two categories are fairly self-explanatory: You look at the situation and decide that saying yes is OK, something you can live with. No bad or stuffed feelings, no hidden agendas, no power imbalance, no confrontation. If you come to a compromise—your way this time, my way next time—you trust that the other person will honor it.

The other two categories are more complex, so let's look at them in depth.

YES—WITH CONDITIONS

When Leigh considered what would make her dinner with Ellen feel less burdensome, she realized that she hadn't given herself any options besides having dinner and spending the rest of the evening with her mother.

I asked Leigh if the world would come to an end if she told her mother that she could have dinner with her but that she'd be going home early.

"Can I really do that?" she asked.

"Of course," I told her. "What you need to tell her is that you've had a rough week and you can go out to dinner but you're not coming over afterwards. And then—and this is really important—you need to say, 'Mom, I really want you to stop comparing me to Caroline every time I say no to you about anything. It really hurts me, and it makes me feel resentful and less likely to do things with you. And I'm warning you right now that I'm going to bust you every time you do it. Deal?'"

Even though this solution was fairly obvious, Leigh hadn't been able to see it. Typically, in our FOG-filled interactions with blackmailers, the obvious eludes us. That's what makes it so important to slow down and observe. Doing so allows you to explore the vast territory that exists right outside the immediate yes that you've become accustomed to giving to your blackmailers. When you become clear about your decision *before* you respond to the blackmailer, you can find compromises that quite frequently satisfy you both.

WHEN THE STAKES ARE HIGHER

When we look carefully at what's being asked of us, we may see that saying yes is indeed a big deal. Although it may not be a major crisis, acceding would violate our standards, our sense of what's right and our self-respect. Even before we know this intellectually, our discomfort level starts to rise and we feel uneasy. Something about the demand goes against our grain. And on some level, we know we can't be complacent.

Like many of us, Zoe was quite good at rationalizing her objections and uneasiness. But when she took the time to look closely at what Tess was asking of her, she realized that the arguments she was using just weren't holding up.

She says she can do it, but I know she wants more responsibility than she can handle. Yet as a friend, and as a boss, I want to give her a chance. That's what's pulling me apart. I don't want to let her down or be uncaring, but I'm worried about putting her on a big account because it's so important to the company. I thought I was just being a perfectionist, but the bottom line is that this is just no place for beginners. I guess that's the thing—these questions about if this is going to hurt anyone. It could hurt me a lot if we can't do good work for this new client, and it could make a lot of other people look bad.

When you evaluate a blackmailer's demand, even a simple question like "Will saying yes hurt me or anyone else?" can help you look beyond the blackmailer's myopic interpretation of a situation. When Zoe did this, she knew she couldn't say yes to Tess without compromising her professional and personal integrity. She'd have to take a stand.

What Money Can't Buy

Jan was mightily tempted by what seemed to be the great bargain her sister offered. She could give Carol a thousand dollars, and in return, she'd be part of the family she wanted so much.

You know, if there was the smallest chance that this loan could bring us closer together, I think it would be worth it. I know it's a long shot given my history with Carol, but maybe she's changed—maybe it'll be OK this time. And I'd be helping her kids. The most I could lose is a thousand dollars—that's not so much.

For Jan, a thousand dollars was a lot of money, but it wouldn't break her to lend it, even if she never saw it again. What she couldn't afford to lose was her integrity. "I have to make up my mind right now, so don't talk to me about integrity," she wailed. "Carol's telling me they're going to be out on the street. No offense, but this touchy-feely stuff is totally irrelevant."

"I'm sure it seems that way when you're under this kind of pressure," I told her, "but indulge me. Just go through this list and see if you still think it's irrelevant."

To help Jan see what a nebulous concept like integrity had to do with her decision about helping Carol, I asked her to answer the following questions. Many people find them helpful when they feel a worrisome buzz around a request but can't pinpoint what's bothering them, or when they want to assess the real cost of a decision.

If I say yes to the request:

- Am I taking a stand for what I believe in?
- Am I letting fear run my life?
- Am I confronting people who have injured me?
- Am I defining who I am rather then being defined by other people?
- Am I keeping the promises I've made to myself?
- Am I protecting my physical and emotional health?
- Am I betraying anyone?
- Am I telling the truth?

You may notice that these questions are based on the components of integrity. They're an effective way to reveal how and where we are not being true to ourselves. Jan found several of the questions sobering.

Am I confronting someone who's injured me? . . . That was like cold water in my face. Because I think of Carol as someone who's really hurt me in the past. She's hurt a lot of people, but no one ever tells her. Then I got to the one about keeping promises to myself. The fact is, after our last big blowout about money, I swore that I wouldn't let her keep pushing me around. She's just not trustworthy when it comes to money. And the most awful question was about telling the truth. Carol hasn't changed, and neither has our family. It's just not real to think that I can wave a magic wand, write Carol a big check and we'll be all warm and fuzzy and happy. Am I betraying anyone if I do it? Yes. Myself.

Jan was silent for a few moments. Then she asked,

How could it be so easy to put all that aside and pretend it never happened? That's more depressing than the fact that I was ready to throw a thousand dollars down the toilet.

When someone asks to borrow money, it usually appears to be a question of whether you can afford to lend it and whether the other person is trustworthy. But money is never just money among intimates. It's a powerful symbol of love, trust, competence, who wins and who loses. Friends and relatives at different levels of achievement or monetary success frequently have seething envies and resentments toward one another that severely contaminate their relationships. It's also very common for people, especially family members, to get set in rigidly defined roles around money: the rescuer, the family hero, the irresponsible and reckless child.

Jan realized that this had happened in her family. Now she was able to make her decision based on new knowledge and awareness. She decided to say no to Carol because she saw that if she gave in to Carol's emotional blackmail, she would be using money to try to buy something that didn't exist. Further, she would be enabling her sister to continue to be reckless with money, which her family had done for years. (I reminded her that this kind of blackmail isn't usually an isolated incident. One request for money usually leads to another.) And most important, she'd have to deny hard-learned truths and break important promises to herself, compromising her self-respect. The cost to her integrity would be much higher than a thousand dollars.

Intimacy with Integrity

The sexual arena is one in which people frequently founder when they run into differing expectations or pressure. There's no place where we're more vulnerable or emotionally naked, and no place where we so want to be accepted—and accepting. If we don't let the other person know what pleases us and what doesn't, what arouses us and what makes us uncomfortable, we can't have genuine intimacy. Yet we don't want to offend or be rigid or closed to playing and experimenting. We know that different people have different levels of comfort and desire, and we want to honor those. We also know the power of sex to attract someone we want, and we know how easy it is to manipulate another person by withholding it. If we're not careful, we wind up making decisions about sex based on all the wrong reasons: To prove we're desirable. To show how free, liberated or spontaneous we are. To assert our claim on the other person. To punish. To escape from FOG.

How do you make decisions in such a sensitive—and murky—area? After all, there are no hard-and-fast rules, just the ones that you and your partner agree to. You need to be clear about what you want and need and to look carefully at what's being asked of you. Then, as in every other part of your life, you need to evaluate the impact of troublesome requests or demands on your integrity and decide what you want to do. It may seem as though questions of a sexual nature are too delicate and nuanced to be analyzed with the same kind of thoroughness we've used so far, but as you'll see, they'll easily survive the scrutiny—and so will you.

Is This about Love?

Sex is all about give and take, and it's fine to do some giving just to please the other person. For example, a man wakes up in the morning ready to make love, and his wife is sleepy and not particularly in the mood but is happy to please him. Nothing's lost, and her integrity isn't at stake, unless this is part of a consistent pattern of man takes, woman complies unexcitedly and without pleasure. In a good relationship between two people who are sexually compatible, going along from time to time won't harm your integrity if it doesn't become obligation or drudgery. Similarly, say a woman asks her lover to indulge a fantasy—"Wear your cowboy boots." It may not be the other person's fantasy, but in a healthy relationship, we ask for pleasure, and we offer it.

But we need to feel free to protect ourselves when what's asked crosses the line and feels harmful to us. Helen talked about how uncomfortable she felt about having sex with Jim one night because she felt she had to regain his affection—even though she was totally exhausted and stressed out. "It was a real low point for me," she said. "I was just so out of it, but he was making me feel so guilty I just went along. I like sex, but this was not fun. I felt used—invisible."

I reminded Helen that there's a big difference between being a good sport, wanting to please the other person even though you'd rather be reading, and being pressured to have sex when you're not feeling well or if you're under a lot of stress. She quickly caught on to that difference. "I love Jim, but I've made my decision," she said. "I'm not going to let this happen again." Helen asked for help in holding her ground, and in the next chapter I'll show you some responses she could give the next time she was faced with this situation.

Bullying someone into having sex when they really don't want to, or don't feel good, is a very unloving thing to do, and a person who's tempted to give in under those circumstances needs to ask: Is this about love, or is it about power, control, winning and domination? If it's about love, the other person will have some compassion for how you're feeling. And if it's not, it's vital to protect your self-respect and your integrity.

MAJOR LIFE DECISIONS: HANDLE WITH CARE

When the stakes involved in a blackmailer's demand are very high, I urge you to stretch out your decision-making process, carefully considering how each option will affect your life and your integrity. I'm talking here about major life decisions such as:

- Deciding the future of a marriage or love relationship
- Breaking off a close relationship with a parent, relative or friend
- Deciding whether to stay in or leave an unhappy job situation
- Spending or investing a significant amount of money

A compromise that preserves the relationship but removes the elements that are unacceptable to you may be perfectly appropriate if the blackmailer is willing to participate. After all, the goal here is not to replace the other person's "My way or the highway" with your own. If you can, you'll want to try to restore the give-and-take flow that emotional blackmail has eliminated.

Give yourself time to explore the blackmailer's demands and the range of responses open to you—except when:

- The other person is physically abusive or threatens to become so
- The other person is compulsive with alcohol, drugs, gambling or debting and refuses to acknowledge the problem or get treatment
- The other person is involved in illegal activities

In these cases, you don't have the luxury of time, and you will have to make your decision and take action quickly.

Holding Patterns: Deciding Not to Decide

Sarah, the court reporter you met in the introduction, had wanted to marry her boyfriend, Frank, but his constant testing of her was making her feel ambivalent. When I took her through the decision-making process, she realized that certain changes needed to occur before she could feel more comfortable about marrying him.

I gave Sarah the assignment of listing what she needed from Frank and what kind of behavior she would and wouldn't accept from him. "Is it OK if I do two lists, the 'You moron, who do you think you are?' list and then the real list?" Sarah asked. "I think I need to blow off some steam."

If you've been stuffing feelings and talking yourself out of your anger, you'll probably want to do the same—or find other safe ways to begin expressing your frustrations—before you focus on your list. Considering what you want and need certainly sounds like a calm, rational process, but the fact is that many emotional blackmail targets have been holding in their resentments for so long that they're close to blowing.

One particularly effective way to vent volatile emotions is to put an empty chair in front of you and picture the other person sitting in it. (A photograph of the person can help you do this.) Say aloud what you've been thinking and feeling for so long. Verbalizing your anger outside the presence of the blackmailer will release bottled-up energy and build your connection to clarity. Yelling or venting with the blackmailer most likely won't clear the air and may instead exacerbate the hard feelings between you.

Sarah began:

I don't know what's happened to us, Frank. You treated me so well at the beginning. I thought I meant a lot to you. But love is not a test. I'm your friend, I'm your lover, maybe I'll be your wife, and I'm outraged that there are so many strings attached to your love. What? We can't get married because I won't baby sit for your sister? How dare you be that petty? How dare you value me on that basis. You can't buy love, Frank, and I refuse to be forced into trying to buy yours. What do you think I am? How can you be such a jerk? Stop it! Just stop it!

Sarah was breathing hard when she finished. She smiled, turned to me and said, "OK. Now I'm ready to do my list."

I told Sarah that in laying out what we want in a relationship, we're not trying to control the situation. We're really saying "This is what would make the relationship more fulfilling to me."

Sarah developed the following list for herself and Frank:

1: No more testing to prove my feelings for you. You either want to marry me or you don't. I love you, and I want to marry you, but I'm no longer going to jump through hoops to prove it to you. If you're that unsure of me, talk to me and we'll work it out.

2: I love you, and I want to expand my business. One does not preclude the other, and the two things can coexist. If they can't in your mind, then there is something fundamentally wrong between us, and we'd better find it out now rather than later.

3: I need you to stop using my resistance to giving you your way about everything as evidence that I'm not committed to you. One has nothing to do with the other.

4: If you want something from me, ask for it, and I'll do everything I can to please you if what you want is OK with me. But I need to be able to say no about some things without you making me feel like a serial killer.

"I feel really good about having done this," Sarah said, "but now I'm worried. What if he just laughs at me? What if he just says no, I can't do that?"

"You won't know until you do it," I told her. "You can rehearse telling him by yourself until you're comfortable, and then you can do it and watch how he reacts. Remember that you're still collecting information. Don't assume anything, but pay close attention. You're making two decisions here. The first is to tell Frank what you need. And the second is to delay making a decision about the relationship until you've seen how Frank responds."

Defusing a Marriage Crisis

Liz had been stuffing her anger for years, and overreacted dramatically when Michael got upset at her desire to go back to work. Both of them had resorted to threats—Liz's to leave the marriage and Michael's to take the twins away from her and leave her penniless. As she looked at what Michael wanted her to do—"Stay home with the kids"—she knew she couldn't do it without giving up something vital to her sense of herself.

I suggested to Liz that she write a letter to Michael, telling him how she felt and once again describing what she needed. If she felt apologies were due, she could make them, and I urged her to use the same kind of nonattacking approach that Sarah used in describing what she wanted from Frank.

Writing a letter to a blackmailer, especially when the situation between you has markedly deteriorated, is a safe way to express yourself. It's a way of keeping yourself from getting so anxious that you forget what you want to say, and it helps you focus on what the key issues are for you. Think of it as a way to find grace under pressure.

Here's what Liz wrote to Michael:

Dear Michael,

I have decided to write my thoughts and feelings instead of trying to express them directly to you for several reasons. The most important one is that I have become very afraid of your anger whenever I try to discuss our situation with you. Now that you have started threatening me with terrible consequences if I decide to divorce you, I am even more afraid. I get all jumbled up in my mind at those times, and I can't think clearly, and I know I don't express myself adequately. For one thing, you constantly interrupt and cut me off when I say something you don't want to hear. By putting what I want to say on paper, I have the chance to organize my thoughts and to state them clearly.

My hope is that you will read this letter through and then we can sit calmly and talk about things rationally and without a win-lose atmosphere.

Michael, I don't want to leave you if there is any chance that we can rebuild our relationship and get it onto a healthier, more loving and equal plane. I still have a lot of love for you despite how much you have hurt me in the last few years, and I know you do for me as well. You can be the most terrific guy in the world (as well as the sexiest), but if I'm going to stay, I need you to take 50 percent of the responsibility for what's gone wrong and do 50 percent of the work to get us back on track.

I promise to do the same. In fact, I'll start right now. I know that I overreacted when you had a cow about my going back to school, and I know that my talking about divorce and seeing a lawyer had a lot to do with you getting so angry and threatening with me. So we both added fuel to the fire and neither one of us talked about how we really felt. I was determined to show you that you couldn't control my life and I take full responsibility for handling it so badly. I'm really sorry.

Until I started seeing Susan I didn't have a term for what was going on between us, but I do now. It is called "emotional blackmail" and it started with us a long time ago. I knew that your "little punishments" like unhooking the garage door opener were insulting and childish, but they seemed minor in comparison to all the good things we had together. I realize now that part of my 50 percent of the responsibility is my failure to let you know how demeaning that behavior was or to let you know that it was not acceptable. Now that the blackmail has escalated to the point of dire threats to try and keep me in line, there must be some major changes or I cannot continue in this marriage.

I am working hard in therapy to rebuild my self-respect, and I am learning a lot about what it was in me that allowed me to accept emotional blackmail for so long. But I can't do it all by myself. I know how much you like getting to the point and problem-solving, so let me tell you what needs to happen from my point of view if we are to have any chance of salvaging what was once a good relationship.

1: The bullying and threats must stop immediately. That is not negotiable. I know that you can't take all the money and the kids, so save your breath! If you are angry with me or frightened about the possibility of my leaving you, you can tell me that, but I will not permit you to treat me like a naughty child, and I will leave the room and the house if necessary if you continue to do so. (Michael, I don't know if you can do that on your own, and I would be ecstatic if you would get some professional help to deal with whatever is causing you to act this way and to learn how to manage your anger.)

2: I want to set aside some private time each evening after the kids are asleep for us to talk to each other respectfully and with kindness. We both have grievances, and I certainly don't expect things to change overnight, but we need to talk them out and find some compromises and solutions.

3: I know that you are much more of a neat freak than I am, and that I do leave things lying around a lot. I will try to pick up after myself, but you need to relax your impossible standards a little, too, and cut me and the kids some slack. Maybe instead of punishing me you could help me.

4: No more yelling. Yelling is an insult to the soul, and besides, it reminds me of my father and scares the hell out of me.

I hope with all my heart that these conditions are acceptable to you. I am more than willing to work with you. Susan has suggested 60 days as a trial period, and that sounds fine to me. Then we can reevaluate things and see how we both feel. Right now I'm feeling very scared but also very hopeful. I think we have a real chance to use this crisis as a wonderful springboard to a better marriage.

—Liz

Michael had been punitive and emotionally abusive, and there was no way of predicting how he'd react to Liz's clear statement of her needs and hopes, but the letter was a positive step for Liz, regardless of the outcome.

Coping on the Job

When emotional blackmail comes up on the job, especially when it involves a superior, it can feel like an insurmountable problem. "Boss from hell" stories are legion, and all the worse because they involve such a great power imbalance. At the back of our minds is the knowledge that our livelihood is in the hands of our blackmailer, and we cede our power to the keeper of the paycheck. Just as in romantic relationships, we may let instances of workplace blackmail pass unremarked on, allowing them to escalate until the only option we feel we have is to leave.

EXPANDING YOUR OPTIONS

Kim, the magazine editor, felt that she was under siege.

I've had it. I spend my life at my desk with my hands surgically attached to my computer and phone. I'm so exhausted I can hardly think straight, and Ken just won't stop with the negative comparisons. I feel like he's holding me up to impossible standards. I'm not a workaholic like some of my coworkers, and if I don't keep cranking at top speed, I'll be working my way off his good list and onto the marginal list—in danger the next time this crazy company decides to downsize again.

There's nothing I can do except look for another job. But I'm worn down physically and emotionally, and all I can do when I get home is try not to burst into tears or yell at somebody who doesn't deserve it. I can't quit, because we need the money. I never used to believe in hell, but I do now.

Clearly, something had to change for Kim. The demands of her job were endangering her physical and mental health, yet she'd made a decision—"There's nothing I can do"—that had cut her options to zero. To get unstuck, she had to make a decision to define for herself what she needed and wanted and then work, even in small, incremental ways, to make changes in her situation.

We began by looking at Ken's demand.

"I don't know how we can do this," she said. "It's not just one demand. It's the whole unending series of them. He thinks I can work all the time, and I can't."

"So what would you say the demand really is?" I asked.

"It's like—Do whatever I say or else."

"Or else what?"

"Or else I'll get rid of you, or at the very least I'll say that you're not as good as Miranda, the greatest editor who ever lived. And once I'm not essential anymore, I'm expendable. Pink slip city."

"We've talked about the negative comparisons with Miranda, but what makes you think that if you don't take on every job that Ken pressures you to take, your job will be in jeopardy?" I asked. "Has he said anything specific like that?"

"Not in so many words," she said, "but it's in the air. Everyone knows that you don't want to fall from grace around there."

"Have you talked to him about the problems you're having with your arms and neck from doing all this work?" I asked.

"Are you kidding?" she said. "We're all just cogs there."

I pointed out to Kim that she seemed to be basing her responses to her boss on some fairly untested assumptions. Then I asked her to describe for herself what it was reasonable for Ken to ask of her.

Once she had a firm idea of what was reasonable, she could turn her attention to what was unreasonable, and look at what the unreasonable elements cost her and other people.

"For people in my job, overtime comes with the territory—this is about a 50-hour-a-week job, plus weekend reading time," Kim said. "I know and accept that—but it's been way more than that. I'm at about 60 or 65 hours, plus going in on weekends. At a really basic level, I hate the pressure. I hate being compared to someone else. It's not motivating me, it's making me feel afraid and resentful."

Finally, I asked Kim to describe what she needed and wanted. "I need to let other people do some of the work I'm doing, and I want Ken to start asking them as well as me," she said. "He's leaning on me too much. I feel so pressured by his negative comparisons that I need them to stop, and I want Ken to just ask me for what he wants instead of twisting my arm that way."

At that point I said to Kim, "You're talking a lot about Ken. What's *your* role in all of this?"

Kim began to think about what she needed to do. "I feel upset that I've let things get this bad. I know I need to learn to say no when I'm tired or hurting or when I need to have a life. It might help if I stop assuming the worst all the time, too."

When Kim thought about the reality of her situation, she could see that much of the pressure she'd felt was coming from inside rather than outside. Was Ken going to fire her if she said she needed to cut back her pace to protect her health? In all likelihood, he wouldn't think of it. She hadn't even mentioned the toll her work was taking on her health—all she'd been able to say to Ken was yes. But she couldn't afford to do that now.

The consequences of allowing herself to work past her limits were far too dire. She decided that what she'd thought was her only choice—to maintain the status quo—was really not an option.

Kim was petrified of approaching Ken, but we practiced what she would say until she felt comfortable. In the next chapter I'll show you how she presented her decision to Ken in a way that enabled them to work far more smoothly together.

CALL IT STRATEGY

If your experience has shown you that you'll probably face unacceptable consequences if you try to approach or resist your boss, then, as long as your physical and mental health are not in jeopardy, you can still choose to go along for the time being.

How do you find a way to work with an employer who is an emotional blackmailer and may be irrational, have a hair-trigger temper or treat you with disdain? Most of us can't and won't change our personalities in order to get by, yet that may seem to be what's asked of us. We know we need to get out of a poisonous situation, but the reality is that without money in the bank or another job offer, many people don't have the luxury of an immediate grand exit.

The answer is to relabel your behavior *strategy* instead of *compliance* or *capitulation*. Doing this will dramatically diminish your feelings of victimization and hopelessness. *Strategy* implies that you've made a choice that's part of a plan that will benefit you, and that's exactly what it should be. Is it devious to look as if you're complying while looking for an escape hatch? No—it's self-preservation.

Guidelines for a strategic holding pattern:

1: Don't tolerate anything that's harmful to your health.

This is one place where it's imperative to protect yourself. You cannot choose to tolerate abuse or imperil your physical or emotional well-being.

2: Decide to define your job to yourself in a different way.

Instead of thinking of your workplace as "the salt mines,"

focus on it as a means to an end that you choose. Tell yourself, for example, "I choose to stay in this situation until I have the financial base I need to make a change." If you're in a position close to the entry level, put your energy into learning all you can, and take advantage of opportunities for formal training or chances to learn from experienced colleagues. Channel the energy of your dislike for the situation into a plan to get out of it.

3: Make a timetable and a plan.

I'm not suggesting that you put up with a difficult job situation indefinitely. What actions will you take to change your situation? Will you look for a new job? Take classes? Get on a promotion track? Change shifts? Save money? How much and how often? Be as specific as you can about what you need from yourself, and make a commitment to your plan.

4: Decide to take small actions to improve your situation.

There's no need to force a dramatic confrontation with an irrational or tyrannical boss, particularly if you're convinced that your job is at stake. But you can take small steps to test the waters and clarify your position. Kim, for example, could interrupt her pattern of always saying yes to Ken by letting him know that she'd made important plans and wouldn't be available at a particular time. She might be surprised to find that he was willing to work with her instead of against her. Some of the biggest bullies will collapse if you hold your ground and begin to stand up for yourself. Paradoxically, they'll also respect you more.

As soon as you decide to derive what benefit you can from a difficult situation, you'll notice your stress level dropping. Remember that you're protecting your integrity by taking care of yourself and making choices that are part of a clear strategy, rather than responding out of fear.

When You Know All You Need to Know

Sometimes enough is enough. We've tried setting limits and expressing our needs to the other person, and we've learned that it just doesn't do any good.

Maria tried for several months to work with Jay on rebuilding their relationship, but to no avail.

You know I've given him every chance, Susan. We've talked and talked, and I asked him to come in for counseling with me, which he did exactly once. And he did agree to go to our minister with me—but he just lied the whole time he was there and charmed him to death.

A relationship is like a pitcher of milk. Sometimes you can get it back in the refrigerator in time, but if it's been left out too long and it goes sour, nothing will ever make it sweet again. I asked Maria if she thought that's what had happened between herself and Jay.

I'm afraid it has, and I can't let him use me like this. Plus the kids will be stuck in this constant tension. I'm ready to snap, and when I look at them, I can see that they are too. Having a mom who's this unhappy is bad enough, but what kind of role model is a father who lies and philanders?

I won't lie to you, Susan. I've looked at this situation every possible way to find a way to keep the family together. It's agonizing for me to have to take this step—I feel like I'm cutting off my arm. But I've realized that in the long run, this is the best thing I can do for the kids too. My life will be better, and in the long run their lives will be better too. When I calm myself down, I can see that the greatest detriment to them would be living with a father like Jay and a bitter, unhappy mother who played the martyr for them. We all need to get this poison out of our lives. It's the only way any of us can heal.

I reassured Maria that from everything I've seen in my work with families, there's no question that she had chosen to do what was best for her children. Parents often believe that they have to stay together "for the sake of the kids," but I've found that it's far more traumatic and destructive for children to be exposed to daily doses of hostility and despair between miserable parents than to experience the clean break of a divorce.

Maria had found the wisdom that would help her begin to find peace. What was left for her now was to stand by her decision.

Standing Up for Your Own Truth

Roberta also concluded that a break was necessary. She couldn't continue to stay in contact with her family.

I need them to accept and believe what I'm telling them—that my father abused me when I was a child. There's no point in my putting down conditions for our relationship, because I've got years of history with these people and I know what they'll do. They won't accept the truth of my childhood, and they say I'm crazy unless I go along with their version. You've seen them for yourself, Susan, and we both know they are closed and joined together in this campaign, and I can't give them what they want, which is agreeing to their version of reality. At least I can't give it to them if I'm going to stay sane. So I guess it's what you always told me—it's them or my mental health. And I'm choosing my mental health.

Roberta decided to let her family know of her decision in a meeting with me at the hospital, a very safe environment. She had the hospital staff, a therapist and the enormous support of her whole environment to help her through a difficult time. When she had presented her decision to her family, she felt lighter, freer and saner, despite their criticisms.

If, like Roberta, you are dealing with abuse issues or have a history of depression or emotional fragility and have made a decision to disconnect, at least for the time being, from certain people in your life, it's important that you have a support system in place. If you don't have a therapist, you will want to enlist the help of people you know are really on your team—a spouse, close friend or sibling. Inform these people of your decision and let them know that you'll be needing their help and support at this critical time of your life.

There are few things more stressful than making a major life decision. Ambivalence, uncertainty, self-doubts and high anxiety are all perfectly normal mental and emotional states at these times. But keep reminding yourself that you are now being proactive instead of reactive. That in itself will help diminish the stress.

Keep using your power statement—"I can stand it"—and continue to visualize yourself moving out of the turbulent realm of emotion and becoming an observer. Both techniques will provide calm and stability during a difficult time. In addition, there are wonderful stress-reducing activities available for all of us. Meditation, yoga classes, dance classes, participating in sports and hobbies, spending time with people you have fun with—all get the endorphins flowing to increase pleasurable feelings and diminish the uncomfortable ones. And, of course, there are many good, inexpensive professional resources that you can take advantage of during this time if you need extra support.

No matter what kind of decision you have to make, use the techniques in this chapter to stop in the midst of pressure, center yourself, and observe what's happening and what's being asked of you. When you make your decision based on criteria that are your own rather than those of the blackmailer, you have dealt a crippling blow to the emotional blackmail cycle. Now let's put your decision into action.

10

Strategy

All the preparation you've done has been leading you here, to that moment when you tell the blackmailer your decision. I know the conflicting emotions going on inside you—the dread, the apprehension and the anxiety that so often accompany behavioral change.

Now I'd like to give you some powerful strategies for stating your case and holding your ground, no matter how the other person reacts. When you practice and use the four core strategies I'll show you in this chapter, I guarantee that you'll change the power balance in the relationship. These strategies—nondefensive communication, making an ally out of an adversary, bartering and using humor—are the most effective tools I know for ending emotional blackmail.

I wish I could stand by your side as you tell the blackmailer your decision, but I can't. What I *can* do is give you a script to learn, hold on to and fall back on when you deal with emotional blackmail.

Please note: If you are living with or involved with someone you feel is volatile and potentially dangerous, do not inform them in advance of your plans to leave them. Protect yourself and get out. If there's a history of physical abuse in the relationship, this is a dangerous time for you. Go someplace safe and get help, if not from family then from a shelter. Don't go alone. Let women's services help you, and take good care of yourself. It would be unrealistic and irresponsible of me to tell you that these strategies will work with a physically abusive person.

STRATEGY 1: NONDEFENSIVE COMMUNICATION

As we've seen, other people have gotten their way by yelling, sulking, playing victim, threatening and blaming. And we've responded by doing the best we could, using the tools we had to work with, to erect a barrier between ourselves and the feelings of fear, obligation and guilt that their behavior stirred up in us.

- We've argued against their descriptions of us. We say: "I am *not* being selfish. *You're* being selfish. How can you say that about me? I do everything for you. How about the time when . . ."
- We've tried to read their minds when they're suffering. We say: "Please tell me what's wrong. What have I done? Come on, tell me what I can do to make you feel better."
- We've tried to buy their approval in the hope that they will stop being upset with us. We say: "Well, if it upsets you that much, I can change my plans/skip my class/not take that job/not see that friend . . ."
- We've tried explaining, contradicting, apologizing, trying to get them to see things our way. We say: "Why can't you be reasonable? Can't you see how wrong you are? What you want is ridiculous/crazy/irrational/insulting . . ."

The problem with responses like these is that they are defensive and actually raise the emotional intensity of the situation. Our attempts to protect ourselves provide the fuel.

What would happen if the sparks of the other person's blame, threats and negative labels fell on damp ground? What would happen if you didn't try to change the other person but instead changed the script? What if you responded to their pressure with statements such as:

- I'm sorry you're upset.
- I can understand how you might see it that way.
- That's interesting.
- Really?
- Yelling/threatening/withdrawing/crying is not going to work anymore and it doesn't resolve anything.
- Let's talk when you're feeling calmer.

And the most nondefensive response of all:

- You're absolutely right [even if you don't mean it].

These phrases are the core of nondefensive communication. Memorize them and add some of your own. Repeat them aloud until they feel comfortable. If you can, practice with a friend. It's important to make these phrases a part of your vocabulary and keep them at the ready. *Do not defend or explain your decision or yourself* in response to pressure.

I know these phrases will feel awkward to you at first. Few of us have experience with answering another person's barrage with a short, unemotional sentence or two. Don't be too concerned if you find yourself tempted to elaborate on them—just don't do it.

Nondefensive communication will work with anyone at any point in the blackmail transaction. I've taught it to thousands of people and used it in my own life for many years. That doesn't mean I found it easy at first, and it certainly doesn't mean that I get it right every time. I had the same butterflies in my stomach and the same pounding heart that almost everyone experiences, and I still get them sometimes. But I promise you that each time you use this strategy and the others I'm going to show you, it gets easier. As many startled blackmailers have discovered, without fuel from the target, the blackmail attempts that worked so well in the past fizzle.

Presenting a Decision Nondefensively

Josh knew that to regain his self-respect, salvage his relationship with Beth and create the possibility of having a real relationship with his father, he had to stop sneaking around and tell him of his plans to marry Beth. I encouraged him to bite the bullet and break the news to both of his parents together, to ensure that his mother would hear his decision from him directly instead of through his father's filters. "I like the idea of using this nondefensive stuff," he said, "but you're going to have to help me because I don't know what to say or how to set things up."

We started with some ground rules for presenting his decision. "First," I told him, "you have to set it up so that you feel as comfortable as possible and the other person can be a receptive listener." In presenting an important decision to another person, you want to give yourself every advantage. That means not trying to initiate a discussion when the other person is tired or stressed out or when the kids are running around the house.

With a husband or partner, let them know you want to talk, and pick a quiet, uninterrupted time. Take the phone off the hook. If you are not living with the other person, say you want to talk, and set a time and a place. Be sure a pick a meeting place where you're at ease. Remember that turf has energy, and it's important not to choose a place that's filled with ghosts of your past, or memories that will make you feel less than equal to the person you're talking to as soon as you walk in the door.

"I can call them and invite them over for coffee and dessert one evening at my place," Josh said, "but I know it's a hassle for them, and there are two of them, just one of me. I think I'll be OK going over there."

I asked Josh if there might be a lot of old memories at his parents' house—pictures or objects that might evoke his childhood. "Oh, no," he said. "It's not where I grew up. They've moved into a condo, and it's more like a hotel than our old house. Look, they're not abusive. They're just being narrow-minded."

Once you've specified a time and place, turn your attention to precisely what you'll say. I suggested to Josh that he start by asking his parents to hear him out without interrupting or contradicting him, and letting them know that when he was done, they could say whatever they wanted. Then he could present his decision. Working together, Josh and I designed the following dialogue.

Dad and Mom, I need you to sit down and hear me out. This is not easy for me to say. I've given this a lot of thought, and because I love and respect you, I want to be honest with you and stop the unpleasantness that's been going on between us. I want you to know that I have made a decision to marry Beth. I'm very ashamed of the fact that I've not been truthful with you about this for the past several months. I've not done it because I'm scared of you. I'm scared of your anger and disapproval. I'm scared to death right now.

Josh accomplishes a lot at the outset. He lays down his conditions for the meeting. He expresses his feelings, both those surrounding the situation and the ones coming up in the present. He acknowledges his earlier dishonesty and his desire to end it. And he announces his decision.

I need you to know that there's nothing you can say or do that's going to change my mind. It's my decision, and it's my life. I'm about to find out whether being right and getting your way is more important to you than having a relationship with me. I hope to God it isn't. I'm sorry I didn't fall in love with a nice Catholic girl. No, dammit, I'm not sorry! And you can either accept this and be part of my new family, or you can decide not to. I love you, Dad and Mom, and I'd suggest that you take some time to decide what you want to do.

Josh continues to stand by his decision, giving his parents a choice of accepting it—or not. Finally, he offers a suggestion: that they not respond instantly but consider what he's said.

Anticipating Their Answers

I encouraged Josh to practice his speech as though he were an actor memorizing his lines. You can do this with another person, talk to an empty chair, or talk to a photo of the other person. It will probably feel strange, but the more you practice, the more confident you'll feel when it comes time to sit down with a real person who has masterfully pressured you in the past.

If you have a number of conditions to present to the other person, it's OK to make notes on a piece of paper and refer to them, letting the other person know that that's what you're doing. But please practice saying your lines aloud—not just in your head. The preparation will give you a tremendous boost.

"I'm happy to practice," Josh said, "but I'm not worried that much about getting through what I have to say. What I'm worried about the most is what *they're* going to say. It'll be bad enough just having to see my father at a slow boil across the table from me."

I helped Josh ease his anxiety about his parents' response by role-playing with him and letting him practice answering the questions and comments he feared the most. Again, you can do this with a friend, or by yourself.

"Which reaction do you think will be the hardest for you, Josh?" I asked.

"From my father I think it'll be 'You know this means I can't help you with your business anymore.'"

"And your answer?"

"Fuck you! I don't need your money!"

"Well, I think we can find something a little less provocative."

"OK. How about 'I'm sorry you feel that way. I've made my decision.'"

We practiced, as you might want to, with a whole series of potential replies.

Susan (as Josh's father): "We just can't support this marriage. I'm hurt and shocked that you lied to me."

Josh: "I'm not proud of lying to you, Dad. I was afraid. I'm sorry you feel the way you do, but I'm going to marry Beth."

Susan: "What will your mom say?"

Josh: "I'll bet the first thing out of her mouth is 'What'll happen when you have children? Will they go to Catholic school? Will you raise them in the church?' We're not even married yet, but Mom's always thinking ahead."

Susan: "And you'll tell her . . . "

Josh: "Mom, we'll raise them with a lot of love and we'll teach them to be good human beings."

Susan (as Mom): "I want to know if they'll be Catholic or Jewish."

Josh: "And I'll say, 'We'll cross that bridge when we have

kids, Mom, when it's a reality. At this moment, that's the last thing I'm worried about.'"

When Josh actually presented his decision to his parents, he felt shaky and intensely nervous, but he stuck to his script, never allowing himself to be put on the defensive.

It wasn't the smoothest encounter in the world. My heart was pounding so hard I was sure they could hear it, and I felt a little sick. I reminded myself to keep breathing, and I repeated "I can stand it" a few times. It helped, but none of this was easy. My father pulled out all the stops. First he said, "Why are you doing this to us? How could you hurt us like this?" I felt like he was stabbing me in the heart, but all I said was, "I'm sorry you see it that way, Dad." He looked surprised, but he just kept going. Next it was "If you marry that girl, you're no longer a member of this family. This is killing your mother." And I said, "Dad, your threats are killing our relationship. I know you're angry, and I know you're upset." Then he actually said something I'd prepared for: "I can't believe you lied to me." My answer was "I did it because I'm scared of you. It's something I hope we can change."

Nothing seemed to be working for him, so he switched to "After everything your mother and I have done for you . . . ," and I said, "Dad, I'm very grateful for everything, but my gratitude does not extend to allowing you to choose who I marry." His last-ditch attempt was to compare me to my brother, who married Catholic and had lots of nice little Catholic kids. I said, "Dad, I can't be like Eric because I'm not Eric, I'm me."

At that point, I saw he was sputtering and running out of things to say, so I did what you suggested. I said that it sounded like he needed time to think.

The last thing Dad said to me was, "You're asking me to handle a lot. I have rules and values and beliefs that mean a lot to me, and I don't know whether I can accept your decision or not." I got up to go and they walked me to my car. I rolled down the window and Dad said, "Well, I always taught you to stand up for yourself, but I didn't mean for you to do it with me." And he kind of smiled, and I drove off.

Josh had faced his worst fear, which was displeasing his parents. And guess what? Nobody died. The building didn't tumble down. The world did not come to an end. It was not an enjoyable experience for him, but it left him feeling relieved and filled with self-respect.

"I feel like I've grown 10 feet!" Josh told me.

He had reclaimed integrity.

In real life, with real people, emotions and interactions are complex, especially in a family; there are rarely Hollywood endings. I'd love to report that Josh's family decided to embrace his new wife, but that wasn't the case. Josh's father decided that he didn't want to lose his son, but so far he has not truly accepted Beth or been loving toward her. Josh has sadly realized that he doesn't want a total break with his parents, but he needs to cut back on the amount of time they spend together because of the tension between them. His fervent hope is that they will soften their attitudes at some point, perhaps when there are grandchildren—and I hope so too. But even if they don't, Josh has done the healthy thing. His self-respect and integrity are intact, and he's far more able to live with himself than he was when he was being untruthful to his parents and betraying his commitment to Beth.

In some instances, parents and others close to us *do* come around. The important thing is what you do about *you*, and who you are when it's time to take a stand.

Handling the Most Common Responses

Because you know the other person so well, you won't have much trouble anticipating the kinds of responses he or she will volley back after you present your decision. But because most of us are so unpracticed in using nondefensive communication, we may not be too quick on the comeback, especially when we try to choose words that will dampen the emotion of the exchange.

Don't worry about being quick—you have all the time you need to think, and it's a good idea to allow a little silence to settle around the other person's words before you speak. The important thing is to resist the tendency to fall back on old patterns because you feel anxious and don't know what to say. So

I'd like to give you some specific ways to answer the most common types of responses. I can't emphasize too strongly how important it is to practice saying these statements until they feel natural to you.

How to respond to the other person's:

1: **Catastrophic predictions and threats.** Punishers and self-punishers may try pressuring you to change your decision by bombarding you with visions of the extreme negative consequences of doing what you've decided to do. It's never easy to resist the fear that their bleak vision will come to pass, especially when the theme they're pounding home is "Bad things will happen—and it'll be your fault." But hold your ground.

When they say:

- If you don't take care of me, I'll wind up in the hospital/on the street/unable to work.
- You'll never see your kids again.
- You'll destroy this family.
- You're not my child anymore.
- I'm cutting you out of my will.
- I'll get sick.
- I can't make it without you.
- I'll make you suffer.
- You'll be sorry.

You say:

- That's your choice.
- I hope you won't do that, but I've made my decision.
- I know you're very angry right now. When you've had a chance to think about this, maybe you'll change your mind.
- Why don't we talk about this again when you're less upset.
- Threats/suffering/tears aren't going to work anymore.
- I'm sorry you're upset.

2: **Name-calling, labeling, negative judgments.** It's the most natural reaction in the world to want to defend yourself when someone starts calling you names, but it's likely to set you up for a pointless "I am not!"/"You are too!" exchange. Instead, take a deep breath and let your feelings of fear, obligation and guilt natter away in your stomach as you move into your head. Remember that for the purposes of presenting and standing by your decision, what's most important is not how you feel but what you say. We are changing behavior first, and later on, we'll turn our attention to what's going on inside you.

When they say:

- I can't believe you're being so selfish. This isn't like you.
- You're only thinking of yourself. You never think about my feelings.
- I really thought you were different from the other women/men I've been with. I guess I was wrong.
- That's the stupidest thing I've ever heard.
- Everyone knows that children are supposed to respect their parents.
- How can you be so disloyal?
- You're just being an idiot.

You say:

- You're entitled to your opinion.
- I'm sure that's how it looks to you.
- That could be.
- You may be right.
- I need to think about this more.
- We'll never get anywhere if you keep insulting me.
- I'm sorry you're upset.

3: **The deadly whys and hows.** The other person may demand explanations from you and want a rationale for your decision. You may be thinking that this is your big chance to say all the things you've wanted to say about how hurt you've been, how thoughtless they've been, how you're mad as hell and you're not going to take it anymore. They're offering you the perfect opening to lay out an elaborate defense. Don't do it!

Stay focused on your purpose. You're presenting your decision—period. Don't get caught up in the content of your disagreement if you want to stop the blackmail process. Your disagreement is really not over where to take the vacation or whether you'll do a favor. It's about a pattern of behavior in which the other person needs to get his or her way and you consistently give in. Because you're serious about breaking that pattern, don't argue, don't explain, don't defend, and don't answer a *why* with a *because*.

Instead, when they say:

- How could you do this to me (after all I've done for you)?
- Why are you ruining my life?
- Why are you being so stubborn/obstinate/selfish?
- What's come over you?
- Why are you acting like this?
- Why do you want to hurt me?
- Why are you making such a big deal out of this?

You say:

- I knew you wouldn't be happy about this, but that's the way it has to be.
- There are no villains here. We just want different things.
- I'm not willing to take more than 50 percent of the responsibility.
- I know how upset/angry/disappointed you are, but it's not negotiable.
- We see things differently.
- I'm sure you see it that way.
- I'm sorry you're upset.

Handling Silence

But what about the person who blackmails through anger that is expressed covertly through sulks and suffering? When they say *nothing*, what can you say or do? For many targets, this silent anger is far more maddening and crazy-making than an overt attack.

Sometimes it seems as if nothing works with this kind of blackmailer, and sometimes nothing does. But you'll have the most success if you stick to the principles of nondefensive communication and stay conscious of the following do's and don'ts.

In dealing with silent blackmailers, *don't*:

- Expect them to take the first step toward resolving the conflict.
- Plead with them to tell you what's wrong.
- Keep after them for a response (which will only make them withdraw more).
- Criticize, analyze or interpret their motives, character or inability to be direct.
- Willingly accept blame for whatever they're upset about to get them into a better mood.
- Allow them to change the subject.
- Get intimidated by the tension and anger in the air.
- Let your frustration cause you to make threats you really don't mean (e.g., "If you don't tell me what's wrong, I'll never speak to you again").
- Assume that if they ultimately apologize, it will be followed by any significant change in their behavior.
- Expect major personality changes, even if they recognize what they're doing and are willing to work on it. Remember: *Behavior can change. Personality styles usually don't.*

Do use the following techniques:

- Remember that you are dealing with people who feel inadequate and powerless and who are afraid of your ability to hurt or abandon them.
- Confront them when they're more able to hear what you have to say. Consider writing a letter. It may feel less threatening to them.
- Reassure them that they can tell you what they're angry about and you will hear them out without retaliating.
- Use tact and diplomacy. This will reassure them that you won't exploit their vulnerabilities and bludgeon them with recriminations.
- Say reassuring things like "I know you're angry right now, and I'll be willing to discuss this with you as soon as you're

ready to talk about it." Then leave them alone. You'll only make them withdraw more if you don't.

- Don't be afraid to tell them that their behavior is upsetting to you, but begin by expressing appreciation. For example: "Dad, I really care about you, and I think you're one of the smartest people I know, but it really bothers me when you clam up every time we disagree about something and just walk away. It's hurting our relationship, and I wonder if you would talk to me about that."
- Stay focused on the issue you're upset about.
- Expect to be attacked when you express a grievance, because they experience your assertion as an attack on *them*.
- Let them know that you know they're angry and what you're willing to do about it. For example: "I'm sorry you're upset because I don't want your folks to stay with us when they're in town, but I'm certainly willing to take the time to find a nice hotel for them and maybe pay for part of their vacation."
- Accept the fact that you will have to make the first move most, if not all, of the time.
- Let some things slide.

These techniques are the only ones that have a chance to interrupt the pattern that's so typical of a silent, angry blackmailer, the cycle that goes "Look how upset I am, and it's all your fault. Now figure out what you did wrong and how you're going to make it up to me." I know how infuriating it is to have to be the rational one when you feel like strangling the other person, but it's the only way I know to create an atmosphere that will allow change to take place. Your hardest job will be to stay nondefensive and to convince the quietly angry person that it's OK for them to be angry when they've spent a lifetime believing just the opposite.

Feeling Furious but Staying Calm

We've talked a lot about dealing with the blackmailer's anger, but how do you stay nondefensive when your own anger is welling up inside you? Allen, whose ex-wife, Beverly, was using their children as pawns in her punitive dealings with him, expressed this frustrating double bind during one session.

I took the kids camping last week, and when I got them back she started screaming because they were dirty and tired. They had a great time, but she said I pushed them too hard. Then she said that if I couldn't take better care of them, she'd see about getting my visitation rights cut down. I know it was a mistake, but I just blew up at her and we started screaming at each other like a couple of lunatics. But she made me furious. How dare she threaten to keep me from seeing my kids? What the hell do I do now?

There are situations for which we have no magic solutions. Beverly had been deeply hurt by the divorce, and since her attacks on Allen had escalated after he remarried, it was obvious that the only way he could change the way she was feeling was to be miserable himself. But he could certainly change whatever he might have been doing to increase the tension.

"I know how furious you are," I said, "but you're just going to have to learn to cool it. You're getting pretty good at using nondefensive communication with Jo, so why don't you try some with Beverly? The hardest thing to do is to act calm while you're feeling homicidal."

"You've got me well trained, Susan," he said with a grin. "I know you're going to say that the only person I can change is myself."

"Right," I answered. "Basically, your job is to zip your lip no matter how irrational she's being and, depending on the situation, to say things like 'I'm sorry you're upset about the camping trip, but they really had a great time. Would you feel better if next time I plan something like this I explain to you before we leave what we'll be doing and what to expect?' You've also told me she doesn't have the boys ready when you come to pick them up and sometimes they're not even there. That's infuriating, but as the custodial parent she has a lot of leverage, and you have to find a way to accept that or you'll be in a constant state of rage and bitterness.

"Again, go back to some of those calm and calming phrases. Instead of ventilating your anger, take a deep breath and say, 'I'd really appreciate it, Beverly, if you'd have the kids ready when I get there. Is there anything I can do to make it easier for you?' I can't predict how she'll respond, but I promise you're going to feel a lot less victimized."

STRATEGY 2: ENLISTING THE BLACKMAILER AS AN ALLY

When emotional blackmail reaches an impasse, it's often helpful to shift the conversation by involving the other person in your problem-solving process. Asking for help, suggestions or information can open up possibilities you hadn't considered, and it's only human nature that other people will be happier to help carry out a decision if they've participated in making it than if they haven't. If you approach the other person with curiosity and a willingness to learn, you can rapidly change the tone of an exchange that's begun to deteriorate into attacks and defenses.

The following questions can help cut through a lot of animosity and tension:

- Can you help me understand why this is so important to you?
- Can you suggest some things we can do to solve the problem?
- Can you help me find some things we can do to make our relationship better?
- Can you help me understand why you are so angry/upset?

In addition, I recommend what I call the Wonder Tool, which sounds like something that ought to be advertised on an infomercial. Actually, it's a strategy of encouraging the other person to imagine with you what change might feel like or how a problem might be solved.

Bring out the Wonder Tool with statements such as:

- I wonder what would happen if . . .
- I wonder if you can help me find a way to . . .
- I wonder how we can do this better/make this work.

To wonder with someone is to unlock the imagination and even a sense of playfulness—nondefensiveness at its most enjoyable. People don't enjoy being attacked, but they're often willing to help someone solve a problem.

Listening for Solutions

Allen's relationship with Jo was far less complex than his relationship with Beverly, since Allen and Jo loved each other and wanted to stay together. But Allen struggled to find a way to deal with his new wife's neediness. After trying for several days to tell her that for the sake of his business, he was going to have to spend some time away from her, he came in to see me, looking for help to find a solution.

I don't know what I can do to keep her from freaking out when I take my trip up north. It's just not going to work for me to say, "I don't care how you feel. I don't care how upset you are. I have to take this trip." Then I'll not only have the trip to worry about but a crying wife to boot.

I told Allen that he might be able to take some of the stress out of presenting his decision to Jo by asking her what would reduce her fears of being left alone. I reminded him that it wasn't his job to fix Jo or deal with the early traumas in her life that had made her so dependent. She had to do that herself so that their marriage could be a partnership rather than a parent-child relationship. Meanwhile, though, he could make her an ally. We practiced how he might use "I wonder" and "I need to understand what I can do" to get Jo to participate in supporting his decision rather than pressuring him to change it.

"OK," Allen said. "How about 'Jo, I need to take a trip to San Francisco for a few days, and before you get upset, I wonder if you can help me understand why you get so neurotic when I step away for two seconds.'"

"No, Allen. We're not trying to label anyone here, just get information. She might have suggestions for improving things, so ask her. How about 'Jo, I have to go up north for a few days on a business trip. I know you get worried when we have to be apart, but this is an important trip, and I wonder what I can do to make you feel better about my going.'"

By putting his dilemma to her in this way, Allen would be acknowledging Jo's feelings. He would not be calling her names, nor would he be leaving the door open to the possibility of his not going.

It was a lot smoother than I thought it would be. I said what we worked on, and as soon as I asked what would make her feel less anxious about the trip she said, "Take me with you." I told her I wouldn't have a problem with that, but I also said that it was going to be a work trip, not a vacation, and she'd probably be by herself a lot since I had so many meetings to go to. At first she said that was fine, she likes hotels, but later she said she'd thought it over, and she'd be more comfortable at home. So it became her choice to stay behind. She just wanted me to call her every evening. God, it was such a relief. We've never worked anything out that way before—it's always been all or nothing.

What had changed was Allen's decision to do what he needed to do and to work with Jo to take her feelings into account. Together, they'd found a mutual solution that Allen might have overlooked or been reluctant to suggest if he hadn't been willing to ally with Jo instead of fighting with her.

Asking the Boss for Help

Kim used a variety of nondefensive techniques to let her boss, Ken, know that she wanted him to stop using negative comparisons and that she needed to cut back her workload to protect her health. She especially liked the idea of enlisting him as an ally because, as she put it,

I'm not in a position to lay down the law and impose my will, but I can do what we're all supposed to do around here—be a good team player. I used to think that just meant do whatever anyone asks, no matter what the cost, but I've begun to think of it more as real teamwork, doing the best I can, pitching in when we're in a crunch and easing off when I need to for my life and my health.

Kim also wanted to put an end to Ken's pressure tactics, and we came up with this way of approaching him:

Ken, you may not be aware of it, but I've noticed that you compare me to Miranda very consistently. In the past, it's been a really effective way to get me to go above and beyond, but it's not going to work anymore. I'll give you 110 percent and do as much as I can without injuring myself, because I want to and because I really like this job. I'm glad you respect me, and I certainly respect you. But please stop playing good kid/bad kid with me. We're two adults. You're not my father, and I'm not your daughter. I'm three years older than you are, for god's sake. And Miranda's not my sister, so I'm busting up this dysfunctional family.

For Kim, as for anyone who is fluent on paper but whose words seem to fail them when they're face-to-face with someone, it was vital to practice. She rounded up a friend to listen and role-play with her, practiced aloud in the car, got help from her husband—and went in knowing her lines cold.

STRATEGY 3: BARTERING

When you want another person to change his or her behavior, and at the same time you acknowledge that you need to make changes of your own, a barter may be in order. Most of us have been making trades since we were kids—two superhero comics for a book, my tuna sandwich for your peanut-butter-and-jelly—giving up something to get something of equal value. The great thing about bartering to reduce emotional blackmail is that it eliminates the idea that the burden of change rests entirely on one person. In bartering, there's no giving without getting. There are no losers.

I saw the power of barter to shake people out of an emotional blackmail impasse when a couple, Matt and Amy, came into my office a few years ago. Amy was furious with Matt for ignoring her.

He treats me like I'm invisible. He gets up, goes to work, comes home for dinner and hardly talks, then sits in front of the television until it's time for bed. He hasn't touched me in weeks, and I have never felt more lonely in my life.

Matt, for his part, said the issue was really Amy's weight:

This is not the woman I married. I think her hobby is eating, and I think you can see that she's gotten pretty large as a result. I just don't find that too appealing. She says I'm acting like I'm not attracted to her and she's right—I'm not. Not when she weighs this much. I'm not going to pretend that it doesn't make any difference to me.

Matt and Amy's relationship had deteriorated to "If you don't find a way to be more loving, I'm going to leave" on her part and "If you don't lose weight, I'm going to continue to punish you by withdrawing" on his part. They didn't put these threats into words, but they didn't need to—their behavior made their feelings as clear as if they'd been shouted through a PA system.

Amy was eating because she felt neglected, and Matt said he was neglecting her because she was eating so much. They were stuck, and blaming each other for their unhappiness. So I set up a barter: Amy would start a diet tomorrow, and Matt would put aside half an hour after he got home each night to talk with her and re-establish contact. Of course, Amy didn't lose weight overnight, and Matt didn't turn into Mr. Communication immediately, but they were able to make significant progress toward breaking out of their stalemate—and, ultimately, mending their relationship.

No one likes to look or feel as though they're giving in, and our distaste for one-sided solutions holds most people back from taking the first step toward resolving a dispute. But bartering creates a win-win situation that's easy for everyone to accept. It also cuts through another dynamic that keeps us from working out problems with someone else—the feeling that they've done us wrong, we're mad about it, and they need to suffer. We won't give an inch because they need to be punished more. But somehow, the sense of getting something from the other person lets us put our resentments aside more easily.

Bartering is a particularly effective strategy because it allows each party to get something they want without the blaming and attacking that is so typical of most conflicts.

Eliminating the Impasse

Bartering allowed Lynn and Jeff to put down the pressure tactics that each had directed at the other. They agreed that at bottom, the unresolved issue in their marriage was the disparity in their financial resources, something that Lynn, in particular, still had problems accepting. But as they sat in my office and talked with each other, they began to see each other as humans again instead of just objects of anger. Each came with a peace offering, and they did their best to be as nondefensive as they could. Lynn began:

I know this money thing is something I need to work on some more. I thought it was OK with me, and we made an agreement when we got together that I wouldn't hold it over you and treat you like a kid getting an allowance or something. So I'm going to honor that agreement. What I need from you, Jeff, is a promise that when something comes up, like buying you a new truck, we'll look at the finances together and make a decision based on what we think we can afford. In other words, no more pressure that you'll disappear if you don't get what you want. I need to understand why you leave without telling me where you're going when you know it makes me so crazy.

Jeff answered:

Sometimes I feel so angry when I have to beg for something I need that I feel like I have to get out of the house or I'll do something I'll regret. I have to let off steam, and when I take off, I don't know how long it'll take to cool down. I don't even know where I'm going half the time.

Lynn responded:

I know how angry my attitude toward money has made you. I apologize for that, and I promise to work on it. I know if we keep talking about this, instead of my holding in my feel-

ings and then taking them out on you, we can resolve the money issue. But I need you to at least tell me you're leaving instead of storming out, and I need you to give me some kind of time frame about when you'll be back. I know you don't always know, but please try. And when you do know, you can call me and let me know where you are and when you'll be back. It'll make me feel a lot better.

Jeff said:

You know I love you and I'm not going anywhere for long. But if it'll help I'll be specific about where I'm going and for how long. And maybe it's time to rethink the finances. I want to look at them together with you—I'm better with money than you think—and I know there are things I can do to make money on the side. I've been thinking about training horses around the valley, but I was so angry at you I didn't even want to mention it. I thought you'd make fun of me because I'm still not going to make as much as you do—probably ever.

Jeff and Lynn still had a lot of talking, listening and negotiating to do, but by using barter they had laid the groundwork for doing that.

Actions, Not Words

Sherry, whose boss and lover, Charles, threatened to fire her when she decided to end their romantic involvement, decided that she'd ask three things of him in a barter that had advantages for both of them: Nonnegotiable was that under no circumstances was she willing to sleep with him any longer. That was a matter of basic integrity. But she did offer to stay at her job until she'd finished the projects she was working on and had helped Charles hire and train her replacement. In return, she wanted an apology from Charles for bullying her and an agreement from him to keep their interactions civil.

I was really afraid that he was going to fire me on the spot, but I'd done a lot of practicing to be sure I knew what I wanted to say, and I think he was surprised that I wasn't scared of him.

At first, he really seemed to be saying "No sex, no job," but when I said I could not compromise on that point, he backed down. He told me, "I don't know if I could handle seeing you every day. I have feelings, too—this hasn't just been a roll in the hay." So I said maybe we could try it out and see how it feels, and he said we could do that. I think it helped that I had something to offer, and that I didn't go in there ready to duke it out with him. I'm working on some things that it would be hard to hand off to a new person, and I think he knew that he'd be better off letting me finish than firing me.

But Charles's behavior did not reflect the agreement he made with Sherry.

Things have really turned difficult. He's been really critical of me in front of clients, and he doesn't miss a chance to get a dig in at me or put me down. He's not keeping up his part of the bargain, and I don't know what to do.

I told Sherry that the only thing to do in her situation was to go back to Charles and let him know that he wasn't doing what he'd promised to. Words aren't enough. They have to be backed up by action. Many emotional blackmailers find it easy to apologize and say they'll make changes but much harder to deliver on their promises. It's important to remind them with words such as "We had an agreement, and I'd really appreciate it if you kept your part of the bargain."

Sherry confronted Charles in a gentle, nondefensive way:

I told him, "You may not even be aware of how hurtful your comments have been, but I'd like them to stop." And of course he didn't have to ask what comments—he knew what I was talking about. And then he half-smiled and said, "You were such a nice person before you went into therapy. . . ."

Even in a case like Sherry's, in which the ultimate goal is to extricate yourself from a difficult situation, it's important to be vigilant and to hold the other person to his or her agreements with you as long as you are with them.

STRATEGY 4: USING HUMOR

In a relationship that's basically good, humor can be an effective tool for pointing out to the other person how their behavior looks to you. Let me give you a couple of examples.

One day when Patty was complaining to me about Joe's suffering, she blurted out: "God, someone should give the man an Oscar: best sufferer in a leading role."

"Why don't you?" I asked her.

She liked the idea so much that she went to a trophy shop and bought a replica of an Oscar. And the next time Joe went into his sulking and sighing routine, she gave him a big smile, applauded and presented him with his award. "'That was brilliant,' I told him. Then I said, 'I especially liked the little sigh at the end.'" The situation suddenly seemed so ridiculous, Patty reported, that both of them burst out laughing—and Joe hasn't been able to suffer effectively since.

Sarah's relationship with Frank was fraying but intact, and she decided that humor might get his attention. She took out an old hula hoop that had been sitting in the coat closet, and the next time Frank set up a condition for their marriage, she said, "Could you hold this while I jump through it?"

"What's this all about?" he asked.

"Well, honey," she said, "I've kind of noticed how you like to keep me jumping through hoops to prove myself to you. Do you think we could talk about it?"

"What are you talking about? I don't do that," Frank said.

"I'm sure you don't realize you do, and I know you love me, but it's felt like a never-ending series of tests to me."

"Hoops, huh?" he said. "OK, let's talk about it."

Then, Sarah reported, "He broke into that grin I just adore and said, 'But before we get serious, do you think you could jump through that little hoop for me first.' It really took the edge off."

There's nothing more intimate than sharing a private joke with someone. Humor is a bond between people, and recalling humorous experiences can be part of the fabric of a strong relationship. Using humor to make a point with a blackmailer can put you both in a relaxed state that lets you remember how much you enjoy each other's company—and gives you a strong reminder of what it feels like to feel comfortable with each other. Humor is healing. It lowers the blood pressure, and it can take the heat right out of a potentially fiery encounter with someone you've been having difficulties with.

If humor is part of your regular vocabulary and you're comfortable with it, it's a wonderful way to express yourself. I can't guarantee that it'll work all the time, but it's bound to make you feel a lot less grim.

EVALUATING THE RESULTS

There's no way of knowing how the other person will respond until you express your feelings and define the limits you need to set in your relationship. As over the years I've worked with targets who came in for consultations with their blackmailers, I've often been surprised by who responded to requests for change. Often people I've expected very little from, because they seemed angry or hard or mean, were actually quite willing to participate in making their relationships stronger. And sometimes those who seemed friendly and flexible turned out to be closed, defensive and not at all sensitive to their targets' needs.

A Positive Outcome

Michael was a dramatic example of someone who responded in a way exactly the opposite of what I had expected. Though Liz had feared an explosion when she presented her conditions to him, she was thrilled by their actual exchange.

I thought a lot about what to do after I wrote that letter. Should I hand it to him and leave the house for a while, or go to his office and drop it off, or just leave it where he could find it? I finally decided that the most comfortable thing for me, because I'm not physically afraid of him, was to sit with him and ask him to hear me out while I read the letter to him.

A couple of times he tried to interrupt, but something really must've touched him because he got very quiet and I could see that he was listening hard. For just a brief moment, I could see the guy I fell in love with sitting across from me instead of that controlling bully. Then his defenses went up and he went after me. He said, "None of this would've happened if you hadn't threatened to get a divorce. Things never would have gotten to this point if you hadn't turned on me like that." I felt like yelling back at him, but I said, "Michael, I'm not willing to take more than 50 percent of the responsibility."

And he calmed down and said, "I guess I didn't want to see that I was hurting you. Why didn't you tell me?" I'm not a Pollyanna, and I know this will take some time to work out, but the most exciting thing is that he's agreed to go to therapy. His hair-trigger temper is a real problem, and I think he sees that his "Me Tarzan, you Jane" routine isn't going to work anymore.

Michael, like a lot of blackmailers, was surprised by how hurt and scared Liz was. I've often heard people who resort to emotional blackmail say "Why didn't she tell me?" or "If I'd only known how much my behavior was hurting him, we could've salvaged things before they got so bad." This isn't prevarication. Blackmailers are often unaware of how painful their behavior and pressure have been, because targets have been too frightened, angry or discouraged to let them know, believing that it wouldn't do any good. In other words, you may not have been saying "Ouch" loudly enough.

We often inhibit ourselves with admonitions like "Don't be a complainer" or "Don't feel sorry for yourself." Some people, men in particular, want to look strong and confident and don't want to seem easily bruised. So we don't express our feelings. We don't say "You're hurting me. Please stop."

Don't be surprised, then, by the other person's surprise at your feelings. No matter what their response, resolve to keep talking, expressing yourself honestly and using nondefensive communication. Then watch to see what the other person does with the new information you've offered.

Sorry Isn't Enough

As I told Liz, after a forthright exchange with the other person we need information that only time can give us. "I know this is a hopeful time for you," I told her, "and I'm excited for you and pleased that Michael is going to go into therapy. I hope this isn't just a honeymoon period, and to be sure things are on the right track, we'll need to keep reevaluating things."

Many times we get very excited by the other person's initial response and believe that a conflict is resolved because he or she has verbally agreed to our terms. But in time we may notice that promises are forgotten and old habits reemerge. We don't want to turn into watchdogs and scorekeepers in our relationships, but we need to look realistically at what's changing and how it fits with what we've decided we want and need.

That's why it's so important to make a decision not to decide on a final course of action until you see what the other person does. When you're making an important decision about the future of a relationship, give the other person time—I suggest 30 to 60 days—and watch for behavioral as well as verbal responses. It's not enough for someone to say, "I'm sorry—now let's not talk about it anymore."

What *is* enough?

1. Accepting responsibility for using fear, obligation and guilt to get their way.
2. An acknowledgment that there are better ways to ask for what they want and a commitment that they will learn those ways.
3. An acknowledgment that their tactics have been unloving and have caused you pain.
4. An agreement to work with you on negotiating a healthier relationship, which may involve getting some sort of outside help if you are unable to work out a problem between you.
5. A willingness to acknowledge your right to think, feel and behave differently than they do, and an agreement that *differently* does not mean "wrong" or "bad."
6. A commitment to work on eliminating the FOG-producing tactics they've used in the past (i.e., no more negative comparisons, no more threats to leave if they don't get their way, no more guilt-peddling and so on).

Changing entrenched behavior—both the blackmailer's and your own—takes time and effort. Give yourself, and the other person, the gift of time.

YOU WILL BE STRONGER

It's scary, telling another person "This is who I am. This is what I want." Scarier still is standing by the truth of ourselves—our integrity—as we must when we give the other person a choice to accept or not accept our decisions and differences. We may feel as though expressing our needs is akin to making demands, but remember that what we're asking of the blackmailer is absolutely reasonable: we want the other person to stop manipulating us. We're not asking for anything that will harm us or them.

Many of us delay presenting our decisions to the other person because we're so afraid of what will happen. But step back for a moment and ask yourself, What's the worst thing that can happen? A common fear is that the relationship will split in two irreparably. But the consequences of not standing up for yourself are worse because *you* will split in two. As time goes on, you'll know less and less of who you are, what you want and what you believe in. Your core will become as thin as a leaf.

If the survival of any relationship depends on your constantly giving in to emotional blackmail, you have to ask yourself if the relationship is worth your well-being. If you get stronger, healthier and more confident and the other person doesn't like it, what does that say about the quality of the relationship that you're trying so hard to save? What is it based on?

In this chapter, we've looked at several relationships that got better and some that ultimately didn't survive. But in each case, blackmail targets pulled free from blackmail to gain a firmer grip on their priceless integrity. No one can predict what will happen as you begin to change, but I promise you that if you use these strategies and face emotional blackmail rather than capitulating to it, you will, no matter what the outcome, be a stronger, healthier person.

11

Cutting Through the FOG

If you have begun to use the tools I gave you in the last chapter, you are developing effective new ways of communicating and behaving. Now I'm going to show you how to disconnect your most important hot buttons.

You may have had some success already in resisting someone's pressure, and you may be seeing changes both in yourself and the relationship. You're tasting the satisfaction and the renewed sense of strength that come from reconnecting with your integrity. But you may have noticed that many of the same old feelings of fear, obligation and guilt that plagued you in the past are still very much with you. It's as though a shiny new building is going up in place of an old one, but the unwelcome tenants who've always lived in the basement won't leave.

There's nothing to be concerned about. Feeling states don't change as quickly as we would like them to. Those feelings have been with you a long time. They took years to become hot buttons, and they're not going to let you evict them without putting up a battle. But it's a battle you're going to win. I'm going to show you direct, practical ways to diminish the lingering hurts and sensitivities that have made you vulnerable to emotional blackmail.

Remember: Even though I illustrate most of these strategies

through my work with other people, all the exercises, role-playing, homework assignments and visualizations are designed for you to do on your own.

OLD FEELINGS, NEW RESPONSES

To those of you who are familiar with my other books, especially *Toxic Parents,* it may come as a surprise that the work we'll be doing here does not in every case involve traveling back to the experiences that are at the root of many of your vulnerabilities. Instead we'll concentrate on changing your responses to those experiences. Certainly we all bear the marks of our pasts. Most of us have at least some awareness of how we were hurt and who hurt us. If we have done some work on ourselves, we have often made the important connection between our emotional scars and the way we behave in our relationships with other people.

But what keeps some of us vulnerable to blackmail is the way we "favor" our injuries. We sabotage ourselves by giving in to blackmail to avoid uncomfortable feelings rather than learning to manage them. It's very much like spraining your ankle and continuing to limp long after you need to because you're afraid of the pain you might feel if you try to walk normally again. I'll be referring to certain childhood experiences, as I have in previous chapters. But what I'd like to help you do here is to learn new responses to old feelings by staying in the present and dealing with the people who evoke those feelings in you *now.*

Note: Before we begin this work I need to reiterate an important point. There are several conditions that mandate some form of professional help. If you are struggling with recurrent depression, crippling anxieties, substance abuse or the aftermath of childhood physical, sexual or emotional abuse, there are many medical, psychological and biochemical interventions that do not require a great investment of time or money. Short-term interactive psychotherapy, the new antidepressant medications, support groups, 12-Step programs and personal growth seminars have changed the face of traditional psychotherapy in the past decade. There is help out there for anyone who truly wants it.

LET'S START WITH THE FEELINGS

Chances are, you know what you do when your hot buttons get pushed. Maybe you're a people pleaser. Or you read about the Atlas syndrome and say, "That's me." Perhaps you avoid anger like the plague. As we begin this vital work of burning off the FOG, I want you to focus on which element or elements of the FOG you're most sensitive to. Take a quick inventory by checking the items below that apply to you.

When I give in to someone who's pressuring me, I do it because:

1. I'm afraid of their disapproval.
2. I'm afraid of their anger.
3. I'm afraid they won't like/love me anymore and may even leave me.
4. I owe it to them.
5. They've done so much for me, I can't say no.
6. It's my duty.
7. I'll feel too guilty if I don't.
8. I'll feel selfish/unloving/greedy/mean if I don't.
9. I won't be a good person if I don't.

You'll notice that the first three statements relate to fear, the second three to obligation and the last three to guilt. It may be that most, or even all, of these statements ring true for you. They did for Eve, who felt afraid of other people's disapproval if she tried to extricate herself from Elliot's smothering dependency, obligated to him because he had given her a place to live and paid her bills, and overwhelmed with guilt at the thought of abandoning him.

In other people, hot buttons tend to be wired to one dominant feeling, although there may be a good deal of overlap among the three feeling states. For example, Liz didn't feel particularly obligated or guilty, but she was afraid of Michael's anger. The statements above will help you figure out which hot button is front and center for you and which element—or elements—of FOG you need to work on for deep and lasting change to take hold.

DISCONNECTING THE FEAR BUTTON

Fear is a basic survival mechanism that's designed to move us out of harm's way. It is both an instinctive and a learned response to danger. If two guys in ski masks are demanding your money, you're going to be afraid. If your husband or wife threatens to take the kids if you leave, you're going to be afraid.

But a lot of the fears we experience in emotional blackmail come up as we anticipate dangers that may or may not exist. The blackmailers in our lives intuitively play on those fears and magnify them. Catastrophic images snowball in our minds, becoming so vivid they seem absolutely real. So we adjust our actions to ward off the emotional blows we're sure are coming. Dealing effectively with fear involves training ourselves to put aside our obsessive worst-case scenarios and develop positive options. You've been letting your imagination work against you. Now let it work *for* you.

Fear of Disapproval

This fear may sound insignificant, but believe me, for many people it is excruciating. The fear of disapproval is much deeper than cringing if someone goes "Tsk-tsk" over something you've said or done. It is interwoven with our basic sense of self-worth. If we allow other people's approval or disapproval to define us, we set ourselves up to believe that there is something fundamentally wrong with us whenever we incur displeasure.

We all love approval and praise from others, and it can sometimes seem like an absolute essential. Many years ago, before I went back to school to become a therapist, I made my living as an actress. I loved it when my efforts were greeted with applause and approval, and I rode the roller coaster to the depths when they weren't. I gauged how well I was doing on the basis of how other people responded to me. But as I've gotten older, I have made a wonderful discovery. I've taken many risks in my life, and I've found that I can tolerate the thundering silence of someone's disapproval or even harsh criticism as long as I stay connected to my integrity.

I know it's not easy to hold on to that connection when people you value tell you you're wrong, but it can be done.

Sarah's relationship with Frank had been improving steadily since she'd made him aware of how much he'd been making her pass little tests to earn the right to marry him.

The talks we're having are helping a lot, but I still can't shake off the idea that I can't feel good about myself or my decisions unless he says they're OK. I keep trying to tell myself to grow up and get over it, but it's not working. I don't want to wind up like my mother, who couldn't even cross the street without getting my father's permission.

A SPECIAL KIND OF COURAGE

Freeing yourself from the fear of disapproval involves knowing which values and judgments belong to you and which have been imposed from the outside. It means knowing what you value about yourself and then having the courage to stand up to disapproval and hold on to your own beliefs and desires.

Sarah was full of excitement when she reported how she had done just that.

You asked me to think about the best parts of me, and what I put at the top of the list were my spirit and my loving a challenge. My business is the way I feed both of those, and I didn't have to think hard to know that I had to move ahead with expanding it. I love Frank, but he's not my whole life. I told him that if he would take some time to think about it, he'd see that I'll be a lot more fun to live with if I'm truly excited about what I'm doing. He mumbled and grumbled a little bit, but I kept using the nondefensive phrases, and he saw I wasn't going to back down. He's fine with it now. I feel like it's Christmas!

Eve's circumstances were different from Sarah's. Sarah had a successful career and a potentially solid relationship. Eve was facing many unknowns—and the job of rebuilding her life—but she too was starting to calm her fear of disapproval.

I've had all these voices coming at me for so long: "You're a cold-hearted bitch." "You're unfeeling." "What a moronic thing to do." "Everything you do is silly." But I'm not going to worry so much about what people think, because what people think in this world can be pretty weird. There are people who think the Holocaust never happened!

On the other side of the fear of disapproval is the freedom to imagine and create a life that really belongs to you. I'm not going to pretend it's easy, but every time you make the commitment to steer your own ship, as Sarah and Eve did, you take a giant step toward shaping a life that reflects what *you* know and what *you* believe is right for you—no matter what anyone else may think or say. When you do that, you'll be able to kick your addiction to approval.

Fear of Anger

Michael made good on his promise to work on his anger, but it didn't take Liz long to see that he wasn't the only one who needed to deal with that emotion. She said:

The other night he tripped on a toy one of the kids had left out, and he started to swear and yell. I was in the other room and he wasn't even yelling at me, but just the sound of his voice made my heart race. He's really been trying to change, and I thought everything would be fine once he got a handle on his anger, but I'm still so sensitive. . . . I don't want to go through life going into a panic every time somebody raises their voice.

Liz wasn't afraid that Michael would physically harm her. Certainly there had been verbal abuse, but she insisted there was never even a hint that he would go any further. So what was it that was creating such a strong visceral reaction in her?

I asked her three questions:

1. What are you afraid of?
2. What's the worst thing that could happen?
3. What's your fantasy of what could happen?

I guess I'm afraid that he'll lose control and run right over me. It's hard to explain. I get this sense of being about two years old and totally helpless. When he's mad, it's like this engulfing, devouring heat. . . .

The sound of Michael's yelling transports Liz through time. She's no longer a 35-year-old adult; she becomes a little girl who hears only danger in a raised voice. That's hardly a surprise given that she grew up in a volatile family where yelling was a cue to duck and take cover. But like many blackmail targets who bend over backward to calm or avoid anger, she kept getting the past mixed up with the present. I told Liz that at some point it would be a good idea to talk to her father and brother about how afraid she used to feel, but for now I wanted her to focus on dealing with Michael's "slips."

No one teaches us how to deal with another person's anger, and most of us have a limited repertoire of responses. The first thing you need to do with a yeller is to choose a calm moment and put them on notice. Say to that person: "I'm not willing to be yelled at, and the next time you yell at me I'm going to walk out of the room." Right away you've taken a strong position and become proactive on your own behalf. Then you need to follow through the next time it happens so the other person will take you seriously.

At the same time you are removing yourself from the conflict, say any of the following in a strong, clear voice: "Stop it!" or "Knock it off!" or, my personal favorite, "Put a sock in it!" Liz looked at me with astonishment. "I can really do that?" she asked.

"Why not?" I told her. "You have my permission."

We often imagine other people's yelling escalating to the point where they lose control and resort to violence. (If you truly fear that another person will harm you, you don't belong in a relationship with them.) But most of us have rarely imagined what would happen if we responded in a more powerful, confident way. Once you step out of the role of fearful little girl or boy and behave like the adult you are now, you're on your way to conquering the fear of anger that triggers so much compliance.

REWRITING HISTORY

An exercise that I have found invaluable for helping blackmail targets handle anger more confidently involves replaying a recent incident when you gave in to someone because you were afraid.

Close your eyes. Rerun the words they said, listen to what you said, and conjure up the memories—the anxiety, the pounding heart, the weak knees, the catastrophic images that ran through your mind as you imagined their anger getting loose and coming at you.

Now play the scene again, but this time, when you see the other person's anger rising, rewrite the scene. Firmly and clearly say, "No. I'm not giving in! Stop pressuring me!" Repeat these phrases until they're convincing—most people start out very tentatively. Listen to how the words sound, and feel how much stronger you are. Yes, you *can* say that, and yes, the words *do* empower you.

Rewrite as many blackmail scenes from your life as you like, as often as you like, unleashing your imagination to experience what it's like to reclaim your personal power. It's especially important to do this exercise if you've been dealing with punishers because they can be so frightening to us. Fear is their modus operandi.

PLAYING THE BLACKMAILER

"One reason I'm so scared of the anger," Liz told me, "is that when I feel it coming at me, it's like the person behind it just disappears. There's no Michael anymore—just this blast of yelling."

I told Liz that I wanted her to step into the role of "The Yeller" and give me an imitation of Michael at his worst.

"You're kidding, right?" she said. "I can't do that."

"Just push past the self-consciousness and give it a try," I said. "Some interesting things might happen. Standing in the blackmailer's shoes for a few minutes can be very revealing."

Liz began hesitantly, but warmed up to the task and launched into an approximation of Michael on the warpath.

If you think you're walking out on me you've got another thing coming. You're not breaking up this family, and if you try, I'm going to make you regret it! You won't get a penny, and I'm not going to let you take the kids! Do you hear me?

When Liz stopped she was very quiet for a moment. Then she said,

It was the strangest thing. I didn't feel powerful at all saying those things. I felt scared and almost helpless, like someone was trying to take something away from me that I really wanted, and the only way I could keep from crying was to yell and scream. I felt like a kid having a tantrum—I didn't have the right words to say, so I just made a lot of noise.

If the blackmailer in your life expresses anger silently, get into the sullen, withdrawn mode and tune into what's going on inside you. See if you can get in touch with how afraid you are of anger and how ineffectual you feel.

Whatever kind of anger you portray, you'll realize that the person you have seen as powerful and in charge is really something of an emotional coward—that's what bullying is all about. Confident, secure people don't need to push others around to get what they want or to prove how strong they are. You may know this intellectually, but "becoming" that person will allow you to experience that awareness physically and emotionally as well.

It is an awareness that can go a long way toward helping you deal with anger, whether you ultimately decide to continue in a relationship with this person or not. Loud, angry punishers and passive-aggressive sulkers are really frightened little kids inside. That doesn't make their behavior any less unacceptable—but it can make it less scary.

Fear of Change

Nobody likes to make major life changes. What's familiar is comfortable, and even though it may be making us miserable, at least we know what's expected of us and what to expect from other people.

Maria was firm in her decision to leave Jay, but she dreaded what lay ahead for her.

I'm afraid, Susan. I'm afraid to be a divorced woman out in the world again. I'm afraid of the pain and the grief. I'm afraid of the uncertainty. I'm afraid of having to start over. I'm afraid I won't know how to make my kids feel safe and secure when it's just me and them. I'm afraid of what people will think—that it was my failure, that I had everything and I blew it. There's such a temptation to call the divorce off and go back to my familiar unhappiness—at least I know how to do that.

Maria could play her assigned role of dutiful wife and mother like a pro, and she knew how to behave in the context of familiar situations. But that comfort, of course, was the problem—it was hard to give up. As soon as we contemplate making any kind of major change, almost all of us feel some degree of panic. And it's that panic that the more destructive blackmailers feed on. So we stay in our old behavior patterns and often cling to harmful relationships and situations to assuage the anxiety and insecurities that threaten to overwhelm us.

I told Maria that I had stayed in a bad marriage for years longer than I should have because I had been afraid of all the same things she was.

"It really helps to know that," she said, "and it helps to know that I'm not some kind of freak or weirdo because I feel this way."

The fear of change is universal, and blackmailers often exploit that fear with statements like:

- You're going to be really lonely without me.
- You'll be sorry, and then it will be too late.

- It's really rough out there for single women.
- How can you do this to the kids?
- You're just not thinking clearly—you don't know what you want.
- Look at all the miserable divorced people out there.

It's perfectly OK to acknowledge your fears to them, but as you do, also reiterate your commitment to change. Say something like "You may be right, and I know it won't be easy, but I'm still going to get a divorce." Or if it's another type of relationship you might say, "I appreciate your concern"—and nothing further. If others persist in painting a gloomy picture of the future to which you've doomed yourself, go back to your nondefensive communication and say, "I choose not to discuss this anymore." Remember: *You have as much right to talk about or not talk about something as they do!*

When you decide to leave or cut off from an important person in your life, you enter a state of crisis—a time of intense emotional upheaval and uncertainty. But crisis is not just a time of danger. When managed thoughtfully and with courage, crisis is also a marvelous opportunity for personal growth and a better life.

This is a good time to find some kind of group or seminar with people in similar situations. To find one, start by asking friends and people you feel comfortable with to recommend programs you know have worked for them. Community colleges and universities have many adult evening programs, and churches and synagogues have formed many support groups that are open to the public. The YWCA or YWHA and women's centers will have a list of resources, and you can call the American Psychological Association for a referral. You don't have to tough it out all by yourself. You need to make sure, however, that the group you've joined is one in which people are working on changing and not just sitting around saying "Ain't it awful" and telling war stories. There is a wonderful healing energy that emanates from people supporting each other through rough times and working toward rebuilding their confidence so that change can become a challenge and not an enemy.

Fear of Abandonment

The fear of abandonment might be the mother of all fears. Some experts believe it's encoded in our genes and is the end-point of all our relationship fears, including the fear of disapproval and the fear of anger. I really don't think it matters whether it's an instinctive or learned fear or a combination of the two. The bottom line is that we all feel it. Some people manage it pretty well, but for others, this fear is profound. When the fear of abandonment causes us to capitulate repeatedly in a self-defeating way, it's as if we were saying "I'll do anything—just don't leave me."

Lynn took great comfort when Jeff agreed not to walk out on her during an argument without letting her know where he was and when he would be back. But her fear of abandonment, which had been part of her for many years, didn't disappear overnight.

> *This really keeps me stuck. If someone gets upset with me, I know they'll wind up leaving me, so I'll just do what they want. I know it's the coward's way out, but I don't care.*

It's a big and not too logical jump from "You get mad at me" to "You'll leave me alone forever," but negative thinking *isn't* logical, and it can easily escalate, turning a simple disagreement into the first step into the abyss.

If, like Lynn, you get pulled into a whirlpool of calamitous thoughts, one of the best ways to escape is to actively limit the amount of time and attention you give them.

THOUGHT-STOPPING

For the next week I'd like you to set aside some time to focus on your negative thoughts of abandonment. Feel free to turn on the doomsday machine and let the fearsome images pour out. But here's the trick: You have to set a timer for five minutes and confine your negative thinking to that period only.

You only need to do this once a day. Think of it as your fret-and-worry time. When your five minutes are up, tell your thoughts, as you would any unwelcome guest, that they need to leave. If they come back during the day, tell them they've had their time, and you'll see them tomorrow. Reduce the time period each day until by the fifth day you're down to one minute. I know this sounds simplistic, but remember: Feelings are triggered by thoughts, no matter how fleeting. We energize our fears by constantly feeding them with mental attention. This thought-stopping technique allows you to interrupt the thought/feeling/behavior continuum at its source and puts you back in the driver's seat.

THE BLACK HOLE

Thought-stopping during the week helped keep Lynn from spiraling into emotional tailspins, but she still had not confronted the fear of what she called "the black hole," the place she would fall into—and never be able to get out of—if Jeff left her. Lynn was not the first person to use that term. I've heard it many times from people who are terrified of abandonment. It seems to be some universally imagined hell.

The black hole had been an image for Lynn as long as she could remember. She was intimately familiar with the dread that surrounded it, and she did not want to cross the threshold and enter. But that, I told her, was what she must do.

"I don't know if I can," she said hesitantly.

"If not today, when?" I asked her. "I want you to take my hand and walk into the black hole with me. What do you see?"

"It's dark, and it's really cold. No human contact. Just isolation with no one to talk to. I'm totally cut off. The days are so long without company . . . The walls close in . . . No one loves me or cares about me or even knows I exist."

Who wouldn't choose capitulation when the only other choice seems to be falling into the bleak, depressed state Lynn describes? And how vulnerable to manipulation you are if you pin your emotional survival on one person!

"OK," I said to Lynn. "You took me in here. Now I want you to find a way out."

"Yeah, right," said Lynn. "I'll just wave my magic wand and the terror will disappear."

"You can get out of here, you know."

"Only Jeff can get me out," she replied.

"No—this one you have to do by yourself or it won't mean anything. I'm not discounting how much Jeff means to you, but he's only one of the elements that enrich your life. Let's start some creative thinking here. What's the opposite of the black hole for you?"

Lynn shut her eyes. "I'm thinking of the other people in my life that I care about—my folks, my friends, some really neat people at work . . . things I love to do—wait, there's a special day I remember. I'm about 12 and my dad got me my first horse—a beautiful palomino. I couldn't believe it! He was all mine. I remember the smell of the hay, the sun on my face . . . I think that's the closest to complete happiness I've ever been."

"And you can go back to that place whenever you feel yourself panicking," I told her. "You can recapture all the sensual pleasures and the excitement anytime you choose. And you can have other days like that. You have a husband and other people who love you, and a good career, and the capacity to feel things deeply. What wonderful gifts! You see—you found your way out of the black hole by yourself!"

The kind of visualization that Lynn did is one we can all do when we feel frightened. Sit down, close your eyes, and take four or five deep breaths. Now remember one of the best days of your life. It might be a day in childhood when you didn't have a care in the world. Or maybe you'll want to return to a beautiful place you visited where all your senses were acute and absorbing the romance and beauty of what surrounded you. Fill up your mind and body with that day, the sights and sounds, the feel of the air, the perfume of flowers or fresh-cut grass. Let yourself fully experience that day until the memories calm you. Remember that you can always use this visualization as a way of bringing light to the black hole.

The fear of abandonment that we feel in love relationships is the adult version of the fear of abandonment we felt as children, when we couldn't survive if we were left alone. Unfortunately, many adults still believe they will face some sort of psychological death if someone they are bonded to leaves them. But the black hole exists only in the imagination. It is a lie masquerading as the truth.

The joyful, precious people and experiences that nourish us tend to fall from our minds when we are frightened. But they're available to all of us both in reality and through our memory and imagination. If fear is like a dark river running through you, you can create stepping stones in the midst of that darkness to help yourself across.

DISCONNECTING THE OBLIGATION BUTTON

I wish somebody could assign us to obligation brackets the way the government assigns us to tax brackets. Wouldn't it simplify our lives if we had a formula for figuring out how much we owe to other people instead of constantly having to wrestle with this dilemma on our own? Wouldn't it be great if there were clear guidelines for knowing how much giving is too much or not enough, figuring out when giving is helpful and when it's harmful or determining how to balance our obligations to others with our very real and vitally important obligations to ourselves?

We are not born with a concept of obligation. We learn it from our parents, in school, from religion and politics and the culture at large. And to complicate matters further, we get bombarded with new rules all the time. For many years, sacrifice and altruism were seen as desirable. Then came the "Me Generation," with its mantra of "Do your own thing." Then the pendulum swung back toward a more compassionate and caring way of relating to others. No wonder we're confused.

It's not easy to sort out where we came by the beliefs about obligation that we've made our own. And in the long run, it doesn't matter. What matters is that you have them—and some of them may be making you vulnerable to blackmail. If you operate from the belief that everyone else's needs are automatically more important than yours, and if you've fallen into a pattern of habitually putting yourself last in any relationship until you exhaust yourself physically, mentally, emotionally, spiritually and financially, it's time to examine—and change—those beliefs.

Where Is It Written?

One of the best ways to start changing some of the beliefs you have about the obligations that are making you feel resentful and stressed out is to see them in black and white. Then you can begin to challenge them.

Start by making a list of what the other person expects from you. Here are some suggestions to get you started:

__________[name] assumes/expects/demands that I will:

- Drop everything to help them
- Come running when they call
- Take care of them physically/emotionally/financially
- Always do what they want during holidays, vacations or leisure time
- Listen to their problems no matter how I'm feeling
- Always bail them out of trouble
- Put my work, interests, friends and activities last
- Never leave them even if they make me miserable

Now, rewrite each statement by preceding it with the phrase "WHERE IS IT WRITTEN" in capital letters. Notice how different "WHERE IS IT WRITTEN that I'm not allowed to enjoy the holidays because I have to be with my husband's family?" sounds, looks and feels from "My husband expects that we'll always go to his family for the holidays." WHERE IS IT WRITTEN that everyone else's needs are more important than yours? WHERE IS IT WRITTEN that you are supposed to sacrifice your well-being to take care of a demanding parent who is perfectly capable of taking care of herself? WHERE IS IT WRITTEN? These seemingly immutable rules, which are keeping you from treating yourself even half as well as you treat other people, are not etched in stone. They exist only in the belief system that is seared into your mind about how you are supposed to be in the world.

Commuting Your Sentence

Karen had a difficult time letting go of her deeply ingrained "I-owe-my-daughter-anything-she-wants-because-she's-had-such-a-hard-time-and-it's-all-my-fault" litany of self-flagellation. She needed to deal with her feelings of obligation on an emotional as well as an intellectual level.

Karen had been judge and jury and sentenced herself to "obligation jail" for a crime she didn't even commit—the car accident that took her husband's life. I asked her to look up the word *accident* in the dictionary in my office.

"It means 'unforeseen, unexpected and . . . ,'" she stopped for a moment and I saw her eyes become teary, "and *unintended!*"

"That's right," I said. "*Unintended.*" I told her to repeat that word often to herself. She hadn't wanted what happened to happen, she hadn't planned it, and she had nothing to do with it. I told her that everyone except some murderers who get life imprisonment without possibility of parole gets out of jail sooner or later. How come she was still there?

I knew that Karen had a rich spiritual life. She attended Al-Anon meetings regularly, often went away on retreats, was an avid student of yoga and meditated daily. Yet Karen had not been able to cross the threshold to self-forgiveness.

I asked her to conjure up a figure that would have the power to let her out of obligation jail, a figure that she could play in a scene with herself. "Well . . . I don't think I'd feel comfortable playing God, but I do believe I have a guardian angel out there—I could play her."

"Great" I said. "You be your guardian angel. Put Karen in the empty chair and get her out of this lousy jail once and for all! And I want you to start by saying 'I forgive you.'"

The tears that had filled Karen's eyes rolled down her face as she began.

I forgive you, Karen. You didn't have anything to do with Pete's death. It was an accident. You've been a good mother, you've been protective and loving to both your kids, you've been a good daughter, a wonderful nurse. You really care about other people—it's time to start giving to yourself. I forgive you, sweetie—I forgive you, I forgive you.

These were things Karen had not been able to say to herself, but in the role of her guardian angel, she could give herself the validation and release she so desperately needed. I urge you to try this exercise. If the guardian angel concept doesn't work for you, you can do it by playing the role of any genuinely loving person in your life. The important thing is to focus on where you got stuck in obligation jail and to get yourself out.

That session was a real turning point for Karen. Toward the end of our hour together she said, "So WHERE IS IT WRITTEN that I'm supposed to dig into my retirement account because my daughter has to have a house right now?"

I told Karen I thought it would be just fine if she wanted to help Melanie financially as long as she was genuinely able to and did it out of love and generosity rather than out of fear of reprisals from her daughter. She acknowledged that the $5,000 that Melanie wanted was too much for her to handle right now, but that she would feel fine with $1,500.

"And if Melanie squawks about it?" I asked her.

Karen smiled and took a deep breath. "Well . . . she's squawked before, and I'm sure she'll squawk again at some future date. I'll just say it's the best I can do and if you want to get mad at anyone, get mad at Susan—she's the one who's responsible for the changes I've made."

People grow up and evolve, but sometimes their beliefs don't change along with them. Like Karen, you are entitled to live by the codes and beliefs that you accept freely as an adult instead of the ones that you assumed automatically and unquestioningly long ago.

How Much Can You Give?

Eve knew that she had to leave Elliot, but all the FOG elements were immobilizing her.

He needs me so much. I do everything for him. And I owe him so much. I just can't get my foot out that door.

This lovely, talented young woman had given up so much of herself to be Elliot's caretaker that she was living on an emotional overdraft with practically nothing in the psychological bank. She had cut herself off from friends, had no fun time or activities that gave her pleasure, had submerged her career aspirations to his and had narrowed her world to a thin sliver.

The more resources you have, the more you can give. It's that simple. If you have a lot of riches in your life—people you love and who love you, emotional and professional satisfaction, friends, fun, enough money—you can probably give a great deal without diminishing your own well-being. Conversely, if you're going through a divorce, having trouble at work and scraping to make ends meet, it will be much harder for you to expend a lot of time and energy to meet someone else's demands. It's a hard lesson to learn, but the truth is that you can't save somebody from drowning if you can barely keep your own head above water.

THE GUILT BUTTON

Guilt draws much of its power over us from the fact that most of us have real difficulty telling appropriate guilt from undeserved guilt. We believe that if we're feeling guilty, it's always because we've done something bad.

Allen's euphoria at being able to talk calmly with Jo and come up with a plan for taking his business trip lasted for about five minutes. Almost immediately, he was squeezed between knowing he was doing the right thing and feeling intensely uncomfortable about making such a huge change in the way he related to his wife.

I know Jo agreed to stay home and didn't seem too upset, but I feel guilty as hell. I get this picture of her in the house alone, curled up on the sofa in front of the TV crying and jumping at every sound she hears. Guilt doesn't come up for no reason, Susan. I'm many things, but I'm not a man who likes to see his wife suffer.

I told Allen that by answering a few questions he would quickly be able to evaluate if his guilt was appropriate or excessive for the situation. I asked him:

- Is what you did or want to do malicious?
- Is what you did or want to do cruel?
- Is what you did or want to do abusive?
- Is what you did or want to do insulting, belittling or demeaning?
- Is what you did or want to do truly harmful to the other person's well-being?

If you answer yes to any of the questions, the guilt you're feeling is appropriate as long as it creates feelings of remorse and not self-hatred. Honoring your integrity will mean taking responsibility for your behavior and making amends. It doesn't mean you're a moral monster.

But if, like Allen, you're doing something healthy for yourself and not trying to harm or diminish another person, then your guilt is undeserved and needs to be confronted. Unless we *do* confront it, excessive guilt can become so entrenched that it becomes like wallpaper—the backdrop for our daily lives.

Allen answered no to all of the questions, but he was still full of ambivalence when he went on his business trip to San Francisco without Jo.

The first night was the hardest. Just as I'd feared, she was crying on the phone when we spoke in the evening. My first instinct was to make a lot of suggestions about what she could do—see friends, go out, visit her family—but I realized that the only way to help her was to stop telling her what to do and let her figure it out for herself. So I told her I missed her, the trip was going well, and I'd talk to her the next night.

The second day was a real turning point for me. When I called her, she wasn't there. I was worried, so I left a message, and when she called back she said she'd been to a movie with her friend Linda. She sounded just fine. It was like—all that worrying for nothing. She was kind of up and down through the week, but she found things to do, and she coped just fine. I'm not saying all this was easy, but we both got by. And it's going to make it a hell of a lot easier the next time.

Use the questions I've listed above to help you whenever your sense your guilt, like Allen's, may be way out of proportion to the event that is triggering it. A healthy conscience will create the amount of guilt that's commensurate with the deed. There *should* be guilt if you sleep with your best friend's husband, and the questions certainly aren't intended to let anyone off the hook for criminal acts. But you don't deserve to feel guilty because you burned the toast or suggested a movie that turns out to be a turkey. And certainly not because you want to do something to enrich your life—even if the other person doesn't like it.

Opinions, Not Facts

The people in our lives who resort to emotional blackmail don't discriminate when it comes to guilt. They will push just as much guilt at us over the small things as the big ones. And oh how willingly we open the door and let the guilt in.

Leigh had told her mother how much the negative comparisons with her cousin had hurt her, and her mother appeared to be receptive. But old habits die hard, and when her mother wanted something that Leigh resisted, she used a different form of pressure.

She wanted me to go down to San Diego with her for the weekend so she could visit my brother and his family, but I have a date and theater tickets, and it just wouldn't work out. I told her she was a big girl and she could get there by herself. I know it was bitchy, but I suggested she go with Caroline. Well, she didn't do the comparison routine, but she did launch into "I guess you're just too busy to spend time with me—you're so wrapped up in your own life you don't care about anyone else. I can't believe you turned out this way!" I know it's manipulative and she's playing the martyr to the hilt, but dammit, I still feel guilty—not as much as before, but more than I want to. I even toyed with the idea of breaking my date and giving my tickets away—but I didn't, so I guess that's progress.

Of course that was progress. Despite the pressure Leigh had changed her behavior, and like so many of us she wasn't giving herself enough credit, because she expected her feelings to change as quickly. One of the things she could do to speed the process of diminishing her undeserved guilt was to learn how to separate her mother's negative labels from fact.

I told Leigh to bring in a list of the most critical things her mother had called her through the years when she was upset with her. I predicted that her labels would probably be familiar to a lot of emotional blackmail targets.

Here are some items from Leigh's list:

- Uncaring
- Selfish
- Thoughtless
- Clumsy
- Pig-headed
- Mean
- Unreasonable
- Rude

Sticks and stones may indeed break our bones, and names like these coming from someone we are close to *can* hurt us. But these labels are not *the truth*. They are someone else's *opinions*. Often we cloak our blackmailers in a mantle of wisdom. They know us better than we know ourselves, we believe, and when they define us in negative ways, we're often quick to buy into these definitions of ourselves, especially if they resonate with what others have told us in the past. In this way we turn another person's opinion into fact. "You are selfish" mutates inside us into "*I* am selfish." It's the same way a child who is told "You are bad" internalizes that message into "*I* am bad."

To help Leigh separate fact from fiction, I had her take her list and opposite each item write the phrase "OPINION, NOT FACT!" in capital letters. Her list then began to look like this:

- Uncaring: OPINION, NOT FACT!
- Selfish: OPINION, NOT FACT!
- Thoughtless: OPINION, NOT FACT!

I'm sure you get the idea. It's important that you absorb this concept.

Sometimes, of course, we *are* uncaring and thoughtless, and it's important to examine the validity of the other person's label. The questions I had Allen answer will help you do that. But most of the time when you're dealing with emotional blackmailers, their name-calling is opinionated, biased and a product of their own self-serving agenda. This is particularly hard to see if the blackmailer is a parent, as in Leigh's case, because we spent our early years believing they were always right. But as you've seen consistently throughout this book, blackmailers are operating out of their own fears and frustrations, and often the very things they accuse you of are the traits and behaviors they evidence themselves. They have *projected* these flaws onto you, expecting you to take them in. Let's send them back.

Return to Sender

The unconscious pays a lot of attention to symbolic rituals and ceremonies. One of the most exciting parts of my work has been creating simple rituals to help clients confront their demons in new and interesting ways. Here's one for guilt that can really help disconnect this hot button.

Find a small box with a lid, like a shoe box. Make it your Guilt Box. Every day for a week jot down the guilt-inducing statements or names that someone uses to pressure you and that you know are unfair and manipulative. Write each one down on a separate piece of paper and put it in the box.

At the end of the week, wrap the box as if you were going to mail it, put the return address of the guilt-peddler(s) on the upper left-hand corner and your own name and address in the center. In large letters, preferably red, write "RETURN TO SENDER" across the front. Then, with as much ceremony as you like, dispose of the box in a way that feels good to you. You can bury it in the backyard, burn it and scatter the ashes, drop it in a Dumpster, put it out with the trash cans or run over it with your car. The point is to stop accepting delivery of the guilt that doesn't really belong to you. It's not yours. Don't let it in.

An Exercise in Paradox

Despite all the demons that were pulling at her, Eve did find the courage to leave Elliot in the most compassionate way possible. She set a specific date when she would be leaving and stayed on long enough to help him find a personal assistant to take over much of the work she had been doing for him. She also alerted his family members to how depressed he was and got several to agree to stay in close touch with him and try to get him to seek professional help.

But I knew Eve was not going to get rid of her mountain of undeserved guilt easily, even though she had made dramatic progress. She had moved in with her mother temporarily, which was working out quite well, and had made some attempts to find a job. But every time Elliot called her and broke down on the phone, she was thrown back into the FOG.

I put an empty chair in front of Eve and asked her to imagine Elliot sitting in it. Then I asked her to get down on her knees in front of the chair and say, "I know you can't make it without me so I will never leave your side. I'm coming back and I will never leave you again. I will give up all my dreams, my aspirations, my very life for you. I will ask for nothing for myself. I will take care of you forever."

Eve looked at me as if I were crazy. "Are you kidding?" she shrieked. "I would never say anything like that!"

"Indulge me," I said.

Eve reluctantly did as I asked. About halfway through she stopped and said, "Wait a minute! I feel ridiculous. I know I'm a soft touch but I'm not a total idiot! I'm not going back! I'm going to have a life! I didn't make him the way he is—why should I have to fix him?"

This kind of work is called paradoxical therapy. A paradox is a contradiction and refers to something that may seem absurd or ridiculous on the surface but that actually turns out to contain some basic truths. Paradox therapy is wonderfully effective. As we saw, Eve's spirit was activated by the absurdity of what I had asked her to say, and she rebelled. Even though she may never have said those actual words to Elliot, until recently her behavior had been saying them for her. Paradox let her take her guilt to a ridiculous extreme and see just how undeserved it was. Once she'd done that, she was on her way to breaking free.

A few weeks later, Eve announced she had found an entry-level job in an advertising agency. She sounded very different from the trapped, hopeless young woman I had first met five months earlier. I asked her if she remembered telling me she was convinced she would "die of guilt" if she ever left Elliot.

"Well, I've never known anyone to die of guilt, and I don't intend to be the first," she said. "I just have to make myself strong and financially independent. I have enough skills to support myself, and all I need is a one-bedroom apartment and a running car. Running water and a running car. I can accomplish that and I'm OK."

She certainly was.

Fighting Guilt with Imagination

Jan was confused when she came to see me after telling her sister that she couldn't lend her money.

> *I know it was the right thing, but I can't shake the idea that I really did something terrible. She's in dire straits. When I think about this, all those old clichés keep coming up: your family is all you have, forgive and forget, blood is thicker than water, the past is the past. Bottom line, she's my sister, she's in trouble and I don't feel good about just leaving her hanging out there.*

Jan was waging an inner battle between what she knew and what she couldn't help hoping for. It was as though the hard-won and painful wisdom that had come from years of dealing with Carol couldn't penetrate deeply enough to challenge her guilt.

When the unconscious is resisting healthy change, I've found that it's effective to reach it through metaphors and stories instead of traditional talk therapy. To help Jan do this, I asked her to write a fairy tale about her relationship with her sister. "That's really going to be a Grimm fairy tale," she said sardonically. "So how do I do this?"

I told her she could write anything she wanted but to use

fairy tale language and images, write the story in the third person, and give it, if not a happy ending, at least a hopeful ending.

The story Jan wrote is very special, and I'd like to share it with you:

Once upon a time there were two little princesses. One was the king's favorite and her closets were filled with beautiful clothes and jewels. She rode around in a golden coach, and she only had to wish for something to have it. The other little princess was the queen's favorite. This princess was smart and brave, but it seemed there was nothing left for her, for her sister had told lies to the king to make the poor princess look bad in his eyes. So the poor little princess wore clothes cast off by her spoiled sister, and when she asked the king for toys or for carrots to feed her pony (she only had a pony instead of a coach), he told her: "Apprentice yourself to a tradesman in town," the king's fancy way of saying "Get a job!" So the poor little princess went to work for the town jeweler, who taught her to make beautiful things, and who praised her talent and industry.

When the princesses grew up, the spoiled princess married a toad who didn't care that she couldn't cook or work. Oh, the toad was handsome enough, but he was a wastrel and a ne'er-do-well. He loved her for her riches, which he wanted to invest in swamp real estate. Soon, all the spoiled princess's jewels were gone, and she and the toad were forced to go begging. This was a great humiliation for the spoiled princess.

In the meantime, the poor little princess had worked hard and made a great success of her life. The kindly town jeweler had let her take over his shop when he got too old to run it, and she was famous for making the kingdom's most beautiful crowns and rings. She had her own jewelry franchise, Princess Jewelry, and she was proud of all she'd worked to make. The only sadness in her life was the memory of how her father and her sister had treated her when she was little.

So when the selfish princess came to her door and pleaded with her to give her some jewels so she could keep the royal carriage from being repossessed and her castle from going into foreclosure, the hardworking princess was faced with a terrible dilemma. "Please help me," the selfish princess pleaded. "I know I've never been nice to you at all, but if you just give me some of what you've worked so hard to earn, I'll be as close as a sister should be."

The hardworking princess wanted to believe her, and she wanted very much to have her sister in her life. But her sister had never been good to her, and the hardworking princess was worried that she hadn't changed at all. To sort out her troubled thoughts, she went for a walk in the woods and came to a crystal-clear pool. She sat there, gazing at her reflection and asking it, "What should I do? What should I do? I know my sister will waste whatever I give her, but how I would love to have a sister's love!" As she spoke, one of her tears fell into the pool, and when the water stilled again, she saw that her reflection had been replaced by the reflection of her best friend.

"You have a sister," the reflection said. "I love you and care for you as your blood sister never did, and you will always have family like me."

The hardworking princess knew that this was so, and when she returned home, she told the spoiled princess, "You can't have the jewels from my shop. You've never had a gift that you didn't lose in the swamp. I wish we could have been close, but we're not, and maybe we'll never be. The jewels can't change that."

Jan said writing the story was very powerful for her:

I really saw the truth. My sister's never going to change. A thousand dollars can't come close to fixing what's wrong. From the time we were little, Carol tried to take things and lied about me and tried to get me in trouble with Mother and Dad. I've never had a close relationship with her, and I probably never will. But I felt so much better when I found myself writing about real sisters. My two closest friends are closer than family, closer than my own sister could ever be. So I haven't lost anything—except a load of guilt.

Putting the story in the third person gave Jan some needed emotional distance, so she could see her relationship with her sister with real clarity. And using the fairy tale form frees up the imagination, with all its creativity and humor—a powerful weapon against guilt. Imagination is as light as guilt is heavy, and it leavens our darkest feelings.

I encourage you to write your own fairy tale to gain a deeper perspective about a relationship that's making you feel guilty. It's especially effective to write about family members, but you can also write about a friend or lover ("Once upon a time there lived a king and a queen. The king used to go into the forest to sulk if he didn't get his way . . . "). I think you'll be surprised and pleased at what your story can reveal to you, and how much clarity it can give you about a situation in which you've been blinded by guilt.

I know I've thrown a lot of information and a lot of work at you in this chapter, and some of it may stir up some strong emotions. You might feel sadness at the loss of safety in a relationship or inevitable anger at a blackmailer for pushing you around and at yourself for giving in so consistently. This work may even activate some unfinished business from your childhood.

Be kind to yourself, and pay attention to your feelings and what they're telling you. If you feel yourself becoming overwhelmed, some brief counseling might be appropriate at this time, or you might want to seek extra support from someone close to you. Remember, you don't have to do everything in the next twenty-four hours. Go at your own pace, and pick those exercises and assignments that apply to you. I promise you that it will be well worth the effort.

Epilogue

Changing your behavior is not a linear process, and it's not instantaneous. As you make the skills you've learned an integral part of your life, you'll find that you won't get it right all the time. You may falter, you may get scared, you'll try and sometimes you'll fail—everyone does. But you'll continue to learn from both your triumphs and your mistakes.

Remember, what you're doing is like climbing a mountain except that no one ever gets to the top. No one is so articulate and anxiety-free that they can always come up with exactly the right words to deflect another person's pressure and threats. Be gentle and forgiving with yourself. As you work your way up this mountain of change, you'll probably glance up and think "Oh god, I've still got so far to go!" But turn around for a moment and look down to the place you started. You'll see how far you've come.

THE MIRACLE OF CHANGE

Once you stop waiting for other people to change and start working on your own behavior, miracles really can happen. Using just one of your new tools will send ripples of change through any relationship. Look at what happened with Liz and Michael.

"Can you believe how different Michael is?" Liz asked me one day. "I really didn't think we were going to make it."

"But who changed first?" I asked.

"I guess I did," she said. "I had my doubts when you said

that's how it works, but I can see that if I'd kept doing what I'd always done, we never would have survived."

Liz smiled broadly as she opened her purse and took out a folded piece of paper. "It's a letter Michael wrote for his therapy, and he asked me to show it to you."

It was quite a letter!

To the blackmailer inside me:

Hello.

I need to have some words with you. I would like your undivided attention to a matter of great importance to me.

You have been the cause of a lot of trouble for me for quite some time now. I didn't really have any idea as to what was going on until Liz and John [Michael's therapist] pointed you out. Many things are much more clear to me now, and you and I are going to have this out right now.

I'm hurting right now because of the tension and unhappiness I have caused because of you. When I think of how close I came to losing everything I love because I stupidly believed that I could feel like a real powerful and in-charge guy by bullying my wife to do whatever I wanted her to do and by punishing her if she didn't, I am appalled and very angry at you.

I am astonished at the scope of my insensitivity. To think that I have looked my wife in the eye and been mean, demeaning and emotionally cruel and thought I was righting some wrong fills me with grief for hurting her, for lost moments, for lost love, for acting exactly the opposite of how I feel, for not respecting and honoring the most important of all things, human dignity and individuality.

I want you to know, Mr. Blackmailer, that there is no place inside of me for your way. I am not willing to compromise about this. It's not OK with me anymore.

I know it won't be easy. There are a lot of things still to learn, a lot of habits to break, a lot of fears of looking weak to overcome. But I have done hard things before that didn't mean as much to me as this, and I will do this, too. Your days are done, and this day and the next are something new.

Good-bye.

—Michael

Like most blackmail targets, Liz had put her faith in complying, believing that she could buy stability by giving in to Michael's demands. She had no way of knowing that she was only reinforcing the behavior in Michael that was tearing them apart. Once she changed her responses to him, she opened the door to the closeness they both craved.

"All I can say is, if this can happen, then I do believe in miracles," Liz said. "I have Michael back—and I have myself back, too."

I can't guarantee that if you do this work you will always be rewarded with a dramatic response like Michael's from the blackmailers in your life. But even if the people around you change very little, *you* will be different and the world will look different to you. You will know that any relationship that can only survive if you remain a compliant target of emotional blackmail is not a relationship that will nourish your well-being.

COMING HOME

A wonderful sense of normalcy and balance returns when you are able to cut through the FOG and interrupt emotional bullying. The confusion and self-reproach that were so much a part of your feelings and self-image lift, and in their place comes a new sense of confidence and self-respect.

With every step you take to learn and use skills that will disarm your emotional blackmailers, you will be restoring the very core of your being—your integrity. That precious wholeness you mourned for was never really lost—just misplaced.

It has been waiting for you.

SUSAN FORWARD maintains offices in Los Angeles, California.
For further information, call (818) 986-1161.

Books by internationally acclaimed therapist

SUSAN FORWARD, Ph.D.

EMOTIONAL BLACKMAIL

(with Donna Frazier)

Emotional blackmail—a powerful form of manipulation in which people close to us threaten to punish us for not doing what they want. Whether they are parents or partners, bosses or coworkers, friends or lovers, emotional blackmailers know our vulnerabilities and deepest secrets, and use this intimate knowledge to gain our eventual compliance. Now in this helpful book, Susan Forward gives blackmail victims the invaluable tools they need to fight back and strengthen their relationships.

"A helpful book . . . [it] makes a whole lot of sense for the person who feels controlled by someone else's needs, whines, and threats."
—Detroit News/Free Press

ISBN 0-06-092897-2 (trade paperback)
ISBN 0-694-51837-9 (audiocassette)

WHEN YOUR LOVER IS A LIAR

(with Donna Frazier)

A powerful book that provides sound advice for women whose men betray their confidence, *When Your Lover is a Liar* shows how to survive and thrive despite deception in relationships. Forward expertly profiles the wide variety of liars, tells how to deal with lies from the benign to the lethal, gives practical strategies to stop them before they ruin your relationship and your life, and shows the path to rebuilding respect and trust in yourself and your partner.

"An outstanding guide for helping women reduce the trauma of . . . interpersonal violations, choose a direction, and rebuild their sense of self."
—Janis Abrams Spring, Ph.D., author of *After the Affair*

ISBN 0-06-093115-9 (trade paperback)
ISBN 0-694-52111-6 (audiocassette)